"Of great value to analysis of land tenure policy in Africa is the book's serious engagement with the contemporary working of African customary tenure, and its pinpointing the social, political and economic conditions in the four country cases where land reform interventions help and where they hinder tenure security for the majority."

Pauline E. Peters, *Harvard University*

"Four rigorous case studies of land tenure reform interventions demonstrate how policy makers often fail to account for the context and complexity of socially inclusive African customary tenure systems. The authors have written a book that should stimulate a reconsideration of land rights policy."

John W. Bruce, *former Director of the Land Tenure Center, University of Wisconsin-Madison*

LAND TENURE REFORM IN SUB-SAHARAN AFRICA

This book examines the impacts of land tenure reform interventions implemented in Benin, Ethiopia, Rwanda, and Zimbabwe.

Since 2000, many African countries have introduced programs aimed at providing smallholder farmers with low-cost certificates for land held under customary tenure. Yet there are many contending views and debates on the impact of these land policies and this book reveals how tenure security, agricultural productivity, and social inclusion were affected by the interventions. It analyses the results of carefully selected, authoritative studies on interventions in Benin, Ethiopia, Rwanda, and Zimbabwe and applies a realist synthesis methodology to explore the socio-political and economic contexts. Drawing on these results, the book argues that inadequate attention paid to the core characteristics of rural social systems obscures the benefits of customary tenure while overlooking the scope for reforms to reduce the gaps in social status among members of customary communities.

This book will be of great interest to students and scholars of land management and use, land and property law, tenure security, agrarian studies, political economy, and sustainable development. It will also appeal to development professionals and policymakers involved in land governance and land policy in Africa.

Steven Lawry is a Senior Research Associate at the Center for International Forestry Research (CIFOR), where he formerly served as Principal Scientist and led the Research Program on Equity, Gender and Tenure.

Rebecca McLain is a Research Program Director at Portland State University, USA, and a Senior Research Associate at the Center for International Forestry Research (CIFOR).

Margaret Rugadya is the Africa Region Coordinator and Senior Program Officer at the Land and Forest Tenure Facility Fund in Stockholm, Sweden.

Gina Alvarado is a Senior Research and Evaluation Advisor at Landesa in the USA.

Tasha Heidenrich is an independent research and evaluation specialist.

Routledge Focus on Environment and Sustainability

Participatory Design and Social Transformation
Images and Narratives of Crisis and Change
John A. Bruce

Collaborating for Climate Equity
Researcher–Practitioner Partnerships in the Americas
Edited by Vivek Shandas and Dana Hellman

Food Deserts and Food Insecurity in the UK
Exploring Social Inequality
Dianna Smith and Claire Thompson

Ecohydrology-Based Landscape Restoration
Theory and Practice
Mulugeta Dadi Belete

Regional Political Ecologies and Environmental Conflicts in India
Edited by Sarmistha Pattanaik and Amrita Sen

Circular Economy and the Law
Bringing Justice into the Frame
Feja Lesniewska and Katrien Steenmans

Land Tenure Reform in Sub-Saharan Africa
Interventions in Benin, Ethiopia, Rwanda, and Zimbabwe
Steven Lawry, Rebecca McLain, Margaret Rugadya, Gina Alvarado, and Tasha Heidenrich

For more information about this series, please visit: www.routledge.com/Routledge-Focus-on-Environment-and-Sustainability/book-series/RFES

Land Tenure Reform in Sub-Saharan Africa

Interventions in Benin, Ethiopia, Rwanda, and Zimbabwe

Steven Lawry, Rebecca McLain, Margaret Rugadya, Gina Alvarado, and Tasha Heidenrich

Routledge
Taylor & Francis Group
LONDON AND NEW YORK

earthscan
from Routledge

First published 2023
by Routledge
4 Park Square, Milton Park, Abingdon, Oxon OX14 4RN

and by Routledge
605 Third Avenue, New York, NY 10158

Routledge is an imprint of the Taylor & Francis Group, an informa business

British Library Cataloguing-in-Publication Data
A catalogue record for this book is available from the British Library

Library of Congress Cataloging-in-Publication Data
Names: Lawry, Steven W., author. | McLain, Rebecca J. (Rebecca Jean), author. | Rugadya, Margaret, author. | Alvarado, Gina E., author. | Heidenrich, Tasha, author.
Title: Land tenure reform in Sub-Saharan Africa : interventions in Benin, Ethiopia, Rwanda, and Zimbabwe / Steven Lawry, Rebecca McLain, Margaret Rugadya, Gina Alvarado, and Tasha Heidenrich.
Description: New York : Routledge, 2023. | Includes bibliographical references and index.
Identifiers: LCCN 2023001002 (print) | LCCN 2023001003 (ebook) | ISBN 9781032430843 (hardback) | ISBN 9781032430935 (paperback) | ISBN 9781003365679 (ebook)
Subjects: LCSH: Land reform—Africa, Sub-Saharan. | Land tenure—Africa, Sub-Saharan. | Farms, Small—Africa, Sub-Saharan.
Classification: LCC HD1333.A357 L397 2023 (print) | LCC HD1333.A357 (ebook) | DDC 333.3167—dc23/eng/20230110
LC record available at https://lccn.loc.gov/2023001002
LC ebook record available at https://lccn.loc.gov/2023001003

ISBN: 978-1-032-43084-3 (hbk)
ISBN: 978-1-032-43093-5 (pbk)
ISBN: 978-1-003-36567-9 (ebk)

DOI: 10.4324/9781003365679

Typeset in Times New Roman
by codeMantra

Contents

Preface

This book studies how contending visions for strengthening land rights in Sub-Saharan Africa play out in the implementation of land tenure reform programs. Formal interventions have primarily taken the form of government land rights certification programs intended to catalyze agricultural investment and productivity and reduce gaps in the legal rights accorded to women and migrant communities. These programs, by shifting authority over land tenure to the state, may run the risk of weakening existing, socially embedded and broadly inclusive and secure customary land tenure arrangements that provide a check on landlessness in rural Sub-Saharan Africa. Despite sharp divisions, we see the potential for new ideas for advancing tenure security to emerge from the interplay of these contesting visions. Among these are interventions designed and targeted in ways that elevate the status of women and others less secure within customary systems, without contributing to the erosion of customary tenure's effectiveness in providing access to rural land as a social right by virtue of membership in the rural community. Chastened at the outset by the complexity of the issues and the political, social, economic, and cultural forces at work, we emerged optimistic at the prospect that African leaders and rural communities will find creative and just solutions to these problems.

This book describes the outcomes of major land tenure reform initiatives undertaken since 2000 in Benin, Ethiopia, Rwanda, and Zimbabwe. Applying realist synthesis methodology, the book elucidates where, how, and for whom tenure interventions increase tenure security, incentivize agricultural investments, enhance agricultural productivity, and support social inclusion with respect to land rights and access, as well as where and for whom they do not provide these benefits. Realist synthesis methodology draws attention to the importance of differences and similarities in social, cultural, economic, and political context in explaining the outcome of interventions.

Funding provided by USAID to Landesa enabled our team to launch the research on which this book is based. In late 2019, Steven Lawry and Rebecca McLain were invited by Landesa to consider leading a realist study evaluating the impacts of major land tenure reform interventions implemented in Sub-Saharan Africa since 2000. At the time, Landesa was supporting USAID in developing a research agenda for land and resource governance issues to better guide USAID's investments in land tenure reform. The report found that the "use of realist synthesis may help USAID and donors more effectively grapple with complexity and the appreciation that LRG (land and resource governance) programs and policies are embedded within complex systems." Gina Alvarado, Tasha Heidenrich, and Margaret Rugadya joined the research team in early 2020.

We are especially grateful to Caleb Stevens at USAID and Benjamin Linkow, then at Landesa, for their leadership in helping get the study underway. We thank Zemen Haddis, Karol Boudreaux, and Janet Nackoney at USAID, and Krista Jacobs at Landesa for comments they offered as the study evolved. We also thank Diana Fletschner, Mercedes Stickler, Jolyne Sanjak, Thom Jayne, John W. Bruce, and Daniel Ayalew Ali for their detailed comments offered at various stages of the research. Pauline E. Peters helped us better understand the social foundations of customary tenure in Sub-Saharan Africa. Ian Scoones and Nelson Marongwe helped us gain useful perspective on the large literature on Zimbabwe's Fast Track Land Reform Program.

Initially, we expected to publish the study as an academic journal article. The final manuscript exceeded the length accepted by most journals, and we are grateful to Hannah Ferguson at Routledge for the interest she expressed in publishing the study as a Routledge Focus Book. Hannah asked three anonymous reviewers to review the manuscript. Their careful and thoughtful criticisms and suggestions helped us to improve the manuscript considerably. Katie Stokes at Routledge oversaw the layout and production of the book. We appreciate her patient assistance as we finalized the manuscript. Lee Greer provided excellent editing of various versions of the manuscript.

We are grateful to the Center for International Forestry Research—International Center for Research on Agroforestry (CIFOR-ICRAF) for providing invaluable research support services. At CIFOR-ICRAF we especially thank Ibu Siswarini Wiwit.

Finally, the views expressed in this book are those of the authors alone, and not USAID, Landesa, or any other individuals or institutions.

The authors

Acronyms and abbreviations

3iE	International Initiative for Impact Evaluation
CFR	Certificat Foncier Rural [Rural Land Certificate] (Benin)
CIP	Crop Intensification Program (Rwanda)
CMO	Context-mechanism-outcome
ELAP	Ethiopia Land Administration Program
FLLC	First-Level Land Certification [Program] (Ethiopia)
FTLRP	Fast-Track Land Reform Program (Zimbabwe)
GPS	Global Positioning System
LAC	Land Use and Administration Committee (Ethiopia)
LRD	Land redistribution
LTR	Land Tenure Regularization [Program] (Rwanda)
LTS	Land tenure security
LUC	Land Use Consolidation Program (Rwanda)
MCA	US Millennium Challenge Account
MRLSL	Matrimonial Regimes, Liberties, and Succession Law (Rwanda)
OLL	Organic Land Law (Rwanda)
PDR	Parcel demarcation and registration
PFR	Plan Foncier Rural (Rural Land Plan) (Benin)
RDC	Rural Development Committee (Zimbabwe)
SLLC	Second-Level Land Certification [Program] (Ethiopia)
SNNP	Southern Nations Nationalities and Peoples Regions (Ethiopia)
SSA	Sub-Saharan Africa
USAID	United States Agency for International Development
VIDCO	Village Development Committee (Zimbabwe)
WBGIL	World Bank Gender and Innovation Lab
WOS	Web of Science

1 Contemporary interventions to reform African customary tenure

More than 965 million women and men in Africa cannot afford a healthy diet (FAO et al. 2020). Many governments and practitioners in the land sector expect that strengthening smallholder productivity will provide sufficient food and enable rural Africans to move out of poverty (Gassner et al. 2019; Lawry et al. 2017). Recent literature examining the link between land tenure security (LTS) and social and economic outcomes suggests that tenure insecurity can inhibit smallholders in Sub-Saharan Africa (SSA) from making land conservation investments such as planting trees, installing soil conservation structures, letting land lie fallow, or other measures that could enhance agricultural productivity, decrease poverty, and improve food security (Higgins et al. 2018; Lawry et al. 2017). LTS interventions in SSA have aimed to create conditions conducive to investment in the farm enterprise, with the twin goals of enhancing agricultural productivity and farm-based incomes (Singirankabo and Ersten 2020; Tseng et al. 2020), as well as reducing food insecurity (Maxwell and Wiebe 1999).

Land tenure security interventions in SSA fall into two major types (Holden and Ghebru 2016): reforms that seek to strengthen existing tenure rights through the provision of state-recognized land certificates, and those that redistribute land, thereby providing landless or land poor members of society with access to an essential means of agricultural production. Beginning in the 1990s and early 2000s, programs aimed at providing smallholders with low-cost certificates for land held under customary tenure were introduced in many African countries (Boone 2019). In politically stable countries, such as Benin and Ghana, an underlying assumption of such programs is that land held under customary tenure without state-sanctioned documentation is insecure (Boone 2019). In countries that have experienced severe political turmoil with associated widespread displacement, such as Ethiopia and Rwanda, reducing land conflicts by providing documentation of land rights was the primary impetus for tenure reforms. A key underlying assumption of land certification programs is that recognition of customary rights through parcel boundary mapping, local rights validation, and locally based rights registration will provide landholders greater tenure security (Boone 2019). In Southern Africa, agricultural reform has focused on redistribution of land,

DOI: 10.4324/9781003365679-1

expropriated by white colonists, to African smallholders who were left landless or with access only to small holdings on marginalized land (Byamugisha 2014). Both customary rights registration and land redistribution (LRD) programs seek to improve the lives of rural inhabitants, either through enhancing tenure security for lands to which they already have access or providing them with secure access to more and better land.

Aside from pointing to the presumed positive impacts of LTS on agricultural productivity, proponents of land tenure reforms argue that tenure institutions can serve important equity needs (Calo 2020; Place 2009). In the context of high levels of poverty and uncertain employment prospects, rural African household members often perceive their customary landholdings, which are secured as a social right, as among their most important economic and social assets (Lawry et al. 2017). There is some evidence of positive outcomes of LTS interventions on social inclusion (Meinzen-Dick et al. 2019), particularly for women, who often own and control less and poorer quality land than men in many countries (Meinzen-Dick et al. 2019). However, the impacts of land certification and LRD interventions on women are variable; often land reforms don't consider discriminatory norms or seek to benefit women in an intentional way. In a cross-country comparison of land rights in SSA, Slavchevska et al. (2021) found that gender gaps were associated with land rights indicators. The persistence of gender gaps in the ownership and control of land in SSA suggests that it is important to understand whether tenure interventions are socially inclusive and contribute to filling such gaps.

In this book, we use a realist synthesis approach to explore the outcomes of tenure interventions aimed at registering customary rights to parcels held by individuals or households in Benin, Ethiopia, Rwanda, and redistributing rights to land in Zimbabwe, as well as the contextual factors that influence those outcomes. We selected these four countries because they were the only ones our systematic database search identified that had sufficient evidence for a realist synthesis. By chance, the four countries encompass diverse regions of the continent, including West, Central, East, and Southern Africa. The four countries also represent a range of colonial histories that have influenced land tenure systems in Africa: French colonialism in the case of Benin, non-colonization and brief Italian occupation for Ethiopia, Belgian and German colonial rule for Rwanda, and British colonialism for Zimbabwe.

Our synthesis seeks to elucidate where, how, and for whom tenure interventions increase tenure security, incentivize agricultural investments, enhance agricultural productivity, and support social inclusion with respect to land rights and access, as well as where and for whom they do not provide these benefits. We argue that inadequate attention paid to the core characteristics of SSA rural social systems obscures the institutional achievements and social benefits of customary tenure while overlooking the scope for carefully crafted reforms to reduce the gaps in social status among members of customary communities. By making visible the underlying program theories of these interventions, a realist synthesis enables us to investigate how they have performed with respect to their desired outcomes and offers revised program theories that reflect more accurately how the programs actually worked in specific contexts and for specific categories of people. Incorporating multiple case studies in the synthesis enables us to identify contextual factors that influence outcomes within and across countries.

Because our primary interest was originally to improve understandings of how, where, and for whom tenure interventions influence agricultural productivity, our study focuses on tenure interventions directed at farmland parcels. Although we acknowledge the importance of commons to rural livelihoods in SSA, common property reforms were not the focus of this research. As a general observation, common property arrangements have not been targets of tenure reform in Sub-Saharan Africa, though Botswana's Tribal Grazing Land Policy (1975) (which granted leasehold rights to large-scale cattle producers on land historically grazed communally under customary tenure) and Kenya's Maasai group ranches program (1982) (granting private title to previously communal grazing areas) stand out as exceptions. In Africa, most rural land used as commons in Africa for forests and grazing is held by the state and managed on the basis of local norms and protocols and in some settings by active state regulation, for instance, by proactive forestry agencies in several Sahelian countries. There are initiatives that seek to devolve forest management rights to local communities, including in Tanzania, Kenya, Madagascar, Mozambique, and some Sahelian states. However, these interventions are quite distinct from our study's focus on tenure interventions aimed at farmland parcels.

Chapter 2 provides an overview of African customary tenure systems and political economy. It compares the key characteristics of customary tenure systems and juxtaposes those characteristics with the program theories that inform parcel demarcation and registration (PDR) and LRD programs respectively in SSA. Before delving into the case studies, in Chapter 3, we describe the methodology we used to develop the four case studies. In Chapters 4, 5, 6, and 7, we identify the contextual factors and theorized mechanisms and outcomes associated with PDR programs in Benin, Ethiopia, and Rwanda, and with Zimbabwe's LRD program. We end each case study with a revised program theory that reflects how the program worked in practice. In Chapter 8, we synthesize findings across the case studies and suggest avenues for future research that can inform the design of LTS interventions. In Chapter 9, we summarize the implications of our findings for LTS intervention programs.

References

Boone, C. (2019). Legal empowerment of the poor through property rights reform: tensions and trade-offs of land registration and titling in sub-Saharan Africa. *The Journal of Development Studies*, *55*(3), 384–400. https://doi.org/10.1080/00220388.2018.1451633.

Byamugisha, F. F. K. (2014). *Introduction and Overview of Agricultural Land Redistribution and Land Administration Case Studies* (F. F. K. Byamugisha, Ed.; pp. 1–16). The World Bank. https://doi.org/10.1596/978-1-4648-0188-4_intro.

Calo, A. (2020). "Who has the power to adapt?" frameworks for resilient agriculture must contend with the power dynamics of land tenure. *Frontiers in Sustainable Food Systems, 4*. https://doi.org/10.3389/fsufs.2020.555270.

FAO, IFAD, UNICEF, WFP and WHO. (2020). *The State of Food Security and Nutrition in the World 2020. Transforming Food Systems for Affordable Healthy Diets*. FAO.

Gassner, A., Harris, D., Mausch, K., Terheggen, A., Lopes, C., Finlayson, R., & Dobie, P. (2019). Poverty eradication and food security through agriculture in Africa: rethinking objectives and entry points. *Outlook on Agriculture*, *48*(4), 309–315. https://doi.org/10.1177/0030727019888513.

Higgins, D., Balint, T., Liversage, H., & Winters, P. (2018). Investigating the impacts of increased rural land tenure security: a systematic review of the evidence. *Journal of Rural Studies*, *61*, 34–62. https://doi.org/10.1016/j.jrurstud.2018.05.001.

Holden, S. T., & Ghebru, H. (2016). Land tenure reforms, tenure security and food security in poor agrarian economies: causal linkages and research gaps. *Global Food Security*, *10*, 21–28. https://doi.org/10.1016/j.gfs.2016.07.002.

Lawry, S., Samii, C., Hall, R., Leopold, A., Hornby, D., & Mtero, F. (2017). The impact of land property rights interventions on investment and agricultural productivity in developing countries: a systematic review. *Journal of Development Effectiveness*, *9*(1), 61–81. https://doi.org/10.1080/19439342.2016.1160947.

Maxwell, D., & Wiebe, K. (1999). Land tenure and food security: exploring dynamic linkages. *Development and Change*, *30*(4), 825–849. https://doi.org/10.1111/1467-7660.00139.

Meinzen-Dick, R., Quisumbing, A., Doss, C., & Theis, S. (2019). Women's land rights as a pathway to poverty reduction: framework and review of available evidence. *Agricultural Systems*, *172*, 72–82. https://doi.org/10.1016/j.agsy.2017.10.009.

Place, F. (2009). Land tenure and agricultural productivity in Africa: a comparative analysis of the economics literature and recent policy strategies and reforms. *World Development*, *37*(8), 1326–1336. https://doi.org/10.1016/j.worlddev.2008.08.020.

Singirankabo, U. A., & Willem Ertsen, M. (2020). Relations between land tenure security and agricultural productivity: exploring the effect of land registration. *Land*, *9*(5), 138. https://doi.org/10.3390/land9050138.

Slavchevska, V., Doss, C. R., de la O Campos, A. P., & Brunelli, C. (2021). Beyond ownership: women's and men's land rights in Sub-Saharan Africa. *Oxford Development Studies*, *49*(1), 2–22. https://doi.org/10.1080/13600818.2020.1818714.

Tseng, T.-W. J., Robinson, B. E., Bellemare, M. F., BenYishay, A., Blackman, A., Boucher, T., Childress, M., Holland, M. B., Kroeger, T., Linkow, B., Diop, M., Naughton, L., Rudel, T., Sanjak, J., Shyamsundar, P., Veit, P., Sunderlin, W., Zhang, W., & Masuda, Y. J. (2020). Influence of land tenure interventions on human well-being and environmental outcomes. *Nature Sustainability*, *4*(3), 242–251. https://doi.org/10.1038/s41893-020-00648-5.

2 Characteristics of African customary tenure and program theories underlying tenure reforms in Benin, Ethiopia, Rwanda, and Zimbabwe

African customary tenure and political economy

Because tenure interventions inevitably transform and reallocate rights (Boone 2019), it is important to understand the characteristics of the tenure systems they seek to replace as well as the political economy in which they are situated.

Chimhowu (2019: 898) defines customary tenure as "an omnibus term that at its most basic means collectively owned land usually under the authority of traditional leadership." The traditional leaders with authority over land include officially recognized traditional chiefs in some areas (e.g., Ghana, Zambia, Zimbabwe); in other areas (e.g., parts of Mali, Burkina Faso, Madagascar), the person with authority over land is the eldest male in the lineage or clan that claims the land. African customary tenure systems are often mistakenly equated with immutable traditional ways of structuring and managing land rights that predate European colonization. However, customary tenure systems are far from static and are subject to both internally and externally driven changes (Cotula 2007). As Boone (2007) notes, colonial and post-colonial regimes modified existing land tenure rules to assert their authority. She refers to these hybrid systems as neocustomary tenure systems (Boone 2015). Chimhowu (2019: 897) uses the term, new customary tenure, to describe these systems, which he describes as retaining many elements of the ideal type of customary tenure found in sub-Saharan Africa, while also having properties, such as formalization and professionalization, which make them visible to market economies. He emphasizes that changes in customary systems come from below through everyday transactions that ultimately transform relationships to land (Chimhowu 2019).

African customary tenure systems vary considerably, depending on the ecological, social, cultural, political, and economic conditions in which they are found, as well as their historical trajectories (Cotula 2007). However, most have some core organizing principles in common, with social embeddedness being the most important (Cousins 2007). Social embeddedness means that membership in the descent group in control of a territory gives one access rights to land. The emphasis on social inclusion and security

DOI: 10.4324/9781003365679-2

distinguishes African customary systems from market-economy models of land tenure (Cousins 2007; Migot-Adholla and Bruce 1994). The key social group in customary landholding systems is the descent group (Peters 2020). In Africa, patrilineal descent, according to which land is inherited through the male line and by sons, dominates. In patrilineal societies where wives come from a different descent group than their husbands (exogamy), wives have use rights to but cannot own their husband's descent group's land (Peters 2019). Although the sisters and daughters within a patrilineal group often have use rights to their descent group's land, inheritance passes through the sons. Although less common, matrilineal-matrilocal descent systems also are found in Africa and affect millions of people (Peters 2019). In these societies, land is passed down to heirs through the female line. Only daughters inherit land, and a husband does not have ownership rights to land belonging to his wife's descent group.

Peters (2019) provides a description of the classic model of how the descent-based tenure system works:

> A classic model is that of nested entitlements with the leader acting as ultimate authority or "trustee/guardian": the term in African languages.…usually connotes ideas of responsibility for the people under the leader's care, along with a variable range of authority. The chief grants access to land of various types to the next level of leaders (e.g., clan or lineage elders) who grant access to constituent families (e.g., compounds or households). Within the latter, persons gain access depending on criteria of social age (usually associated with marriage), gender, and need.
>
> (pp. 49–50)

The lineages that first settled in an area typically have the strongest land rights (Migot-Adholla and Bruce 1994). Newcomers can acquire membership in the group through marriage or alliances, and so acquire either direct or indirect rights to land (Krantz 2015). Members of the social group have a right to some land for support and usually, but not always, security of tenure in particular parcels of land (Migot-Adholla and Bruce 1994). However, rights to land come with responsibilities and obligations toward others' rights (Cousins 2005), and mutual interdependence (Peters 2019).

Customary tenure systems are inclusive in the sense that individuals and households who belong to the social group (often a lineage or clan) are considered to have legitimate claims to lands over which that group exerts control. However, Platteau (1996) and Peters (2009) note that distinctions with respect to land claims have long been made in many African tenure systems between those who are perceived to belong to the social group (autochthons) and those who are not (strangers). Peters (2013) concludes that exclusion and inequality are associated with embeddedness, as well as inclusion.

The importance of kinship, marriage, and descent practices in African land tenure systems cannot be overemphasized. Peters (2021) argues that to understand land dynamics in African societies, it is imperative to understand the interactions between kinship and descent practices and social differences, such as gender, age, wealth, and class, among others. As an example, she critiques the notion that customary rules in patrilineal societies that require land to pass to a man's sons rather than to his wife upon his death are motivated by a bias against women. She argues that the purpose of these rules is to ensure that the husband's lineage retains control over their land, rather than discrimination against women (Peters 2021).

Honig (2017) uses the term, "customary privilege" to characterize the link between social status and tenure security. She argues that individuals or households with high customary privilege are more likely to enjoy greater tenure security. The determinants of social status vary by society, but landholders who are descendants of the families that first cleared the land in an area typically have higher status than those whose families settled in the area later (Honig 2017). Migrants, even if they have been present for many generations, and particularly those from an ethnic group different than that of the founding families, often have weaker land rights under customary systems (Boone 2007; Honig 2017). Tensions over land rights between founding lineages and migrants have periodically contributed to violent political conflicts in countries such as Ivory Coast, Liberia, and Eastern Democratic Republic of Congo (Bøås 2009).

Women in African societies often have weaker land rights in relation to those held by men (Tsikata 2016). However, it is important to differentiate between different categories of women. A key difference is the woman's descent group. As Peters (2021) notes, the land entitlements that daughters and sisters have in patrilineal societies differ from those of wives and widows. This is because the daughters and sisters are members of the descent group, whereas the in-marrying wives are not. In the less common situation involving societies with matrilineal-matrilocal descent groups, it is the husband who is unable to own his wife's land, with the logic being the same as for patrilineal societies: land should remain in the descent group (Peters 2021).

Other factors besides kinship and descent practices influence the strength of women's rights to land. The relative economic contribution a woman makes to the household can also make a difference in the access a woman has to land. Tsikata (2016) found that women who made major self-provisioning contributions to the household were more successful at gaining rights to farmland that they needed to meet their production obligations. Mutopo (2011) describes an example from Zimbabwe where women negotiated with their husbands for more land for their own use after resettlement on larger parcels made available by the Fast Track Land Reform Program. Among the Antakarana in Madagascar, Gezon (2002) found that older women with high social status often played a strong role—albeit behind the scenes—in land use decisions. Indeed, the flexibility and room for negotiation that characterizes many customary tenure systems in Africa has been cited as a reason for promoting land reforms that recognize customary rights (Peters 2009). However, flexibility may not always be advantageous, particularly for lower-status community members who may lack the social capital or skills to negotiate effectively with more powerful, higher status community members (Peters 2009).

Flexible boundaries are another feature commonly found in African customary tenure systems, particularly in areas where extensive livestock production is an important component of livelihood strategies, and mobility is a key adaptation to highly variable rainfall regimes (Turner et al. 2016). The presence of flexible boundaries represents a central challenge to the mainstream cadastral model of land registration or titling, which emphasizes the establishment of immutable boundaries (Lavigne Delville and Moalic 2019).

The presence of nested systems of overlapping rights to land and natural resources is another characteristic of African customary tenure systems (Peters 2021). These overlapping claims are "embedded in institutions – lineage, familial, neocustomary, chieftaincy, patriarchal, age-sets, and so forth" (Boone 2019: 394). Benjaminsen (2002) gives an example from Mali, where one parcel of land is used as farmland by one individual, as a fruit harvesting site by another, and by yet a third as a place to let livestock forage after the crops have been harvested. Moreover, because African customary systems typically have both group and individual entitlements and associated responsibilities, they simultaneously have elements of communalism and individualism (Cousins 2007). In most rural African contexts, the system of land rights includes resources that are managed as common property, such as grazing lands, woodlands, and water, as well as farmland held collectively by extended families and lineages or individually (Okoth-Ogendo 2002). Land reforms that disrupt the web of social relations associated with overlapping and nested land claims are likely to not only reduce tenure security but also are likely to disrupt the web of social relations that are essential to the livelihoods of many rural Africans (Peters 2020).

The political economy context of tenure security interventions in Africa

Cousins (2005) asserts that land rights in customary tenure systems are politically, as well as socially embedded, and that power relations affect how rights and benefits are distributed. Boone (2007) concurs but argues that tenure reforms do much more than redefine rights, they also reconfigure how communities interact with each other and how communities and the state interact. Boone (2007, 2015) groups the land regimes that have resulted from colonial and post-colonial tenure reforms into two types. One type, which Boone labels neotraditional communal tenure (or neocustomary land regimes), consists of systems established during colonial times in which the colonial regime (and subsequently post-colonial state) recognized de facto or de jure neotraditional local authorities as having land management powers. When the colonial powers established these systems, indigeneity (or non-indigeneity) governed whether one had access to land and the conditions under which one had access. Members of the descent-based community were automatically entitled to have access to land, whereas strangers had no such entitlement.

Boone (2007, 2015) labels the second type of land regime as a statist, or user rights regime. In statist regimes, the colonial regime challenged the existing rights of local authorities to land, and instead supported the claims of individuals or households using the land. Post-colonialist states have applied statist land regimes through initiatives such as resettlement schemes in Ethiopia and marshland reclamation in Rwanda (Boone 2015). Under the statist scenario, the state takes a direct role in land allocation and adjudication, and land users are beholden to the state for access to land rather than to neotraditional authorities (Boone 2015).

Both types of regimes can (and do) exist within a country and may be in direct competition and conflict within specific areas. For instance, Mkodzongi (2016) demonstrates how conflicts arose between chiefs and the state over land allocation authorities in Zimbabwe's FTLRP resettlement areas. Although resettlement areas are legally designated as state land under the administrative authority of the Lands Department and its local representatives, chiefs have become key players on rural land matters by deploying claims of ancestral autochthony over newly resettled farms (Mkodzongi 2016). Locally, chiefs may have greater political legitimacy than local state officials and, in many settings, may be able to deliver administrative services more efficiently. But the law provides that the Land Department formalizes transfer of land to widows on the death of their husbands, and not traditional authorities (whose commitment to women's land ownership may in some settings be questionable). Here, the competition between neotraditional and statist claims of authority can create considerable uncertainty for landholders.

With respect to tenure reforms, Boone (2007) points out that supporting the strengthening of community rights where neocustomary regimes are present risks legitimizing and strengthening rules and norms governing access to land that discriminate on the basis of social status (i.e., gender, age, ethnicity, indigeneity). Supporting or promoting statist land regimes, on the other hand, expropriates land from the original claimants in favor of those using the land, who may or may not be well-off. Moreover, because multiple and overlapping rights prevail in much of Africa, tenure reforms that purposefully or unintentionally promote the individualization of landholdings effectively expropriate that land from all other claimants (e.g., those not granted individual rights) (Boone 2007). As Boone notes, who wins and who loses varies depending on the type of land regime present in an area: members of extended families and communities lose out in areas where communal controls were dominant whereas in areas where user rights were enforced, the indigenous landholders and their descendants lose out. In both cases, the individualization of land results in a fundamental transformation of the social and political order over the long term (Boone 2007).

Peters (2013) identifies a number of contextual factors that, by increasing land values, threaten to undermine African customary tenure systems and transform the social fabric in ways that are likely to increase inequities in access to land and other resources. Among the primary threats are an increase in cash crop production, greater demand for arable land and residential land in response to population growth, and the appropriation of land for conservation and resource extraction. To this list, Cousins (2007) adds weak state administration, abuses by traditional authorities, and tensions between traditional and local state authorities over their respective roles and responsibilities. Moyo (2007) describes an increase in land concentration due in part to internal social differentiation and in part to urban and rural elites and foreign investors working through the state administration to acquire land. Jayne et al. (2019) document an expansion in middle-sized farms in several African countries. They describe the impact of the increase in middle-sized farms as mixed for smallholders. On the one hand, they compete with smallholders for increasingly scarce land but at the same time, the emergence of medium-sized farms has increased nearby smallholders' access to services and markets. Despite these threats, Cousins (2007) argues that in many areas, existing customary regimes are relatively stable such that the majority of landholders perceive their tenure to be secure.

Research by Lawry et al. (2017) and Childress et al. (2018) likewise suggests that under some circumstances, customary tenure systems provide greater tenure security than they have been given credit for in the past. In the context of weak states, high levels of poverty and uncertain employment prospects, many rural African household members continue to perceive their customary landholdings, secured as a social right, as among their most important economic and social assets (Lawry et al. 2017). Socially inclusive tenure systems by design accommodate the needs and interests of a great variety of household types, albeit differentially depending on the household's endowments of social, personal, financial, and political capital. The contribution of agriculture to household income will vary based on trade-offs that families and individuals make about how best to employ the talents and assets at their disposal (Migot-Adholla and Bruce 1994).

Program theories shaping the design of tenure reform interventions

The parcel demarcation/mapping and rights registration systems implemented in Benin, Ethiopia, and Rwanda were designed to provide rural landholders with clearly demarcated and publicly adjudicated boundaries for their land; a map of their parcel and a description of the rights and rights holders associated with the land; and, ultimately, a document certifying that information about the parcel and the persons with rights to it had been registered at the local land administration office. In Rwanda and Ethiopia, registration with local land offices was included as part of the intervention; in Benin, the PDR program trained commune land administrators to issue land certificates, but certificate issuance was not an integral component of the program (WBGIL 2019). Nonetheless, the intention of the intervention in Benin was that landholders would seek a state-issued land certificate from the commune land office once their parcels were mapped and rights registered (WBGIL 2019).

Although the PDR programs for Benin, Ethiopia, and Rwanda differed somewhat in their design, all three programs were based on the premise that customary land tenure institutions fall short in providing adequate levels of tenure security needed to promote agricultural investment and productivity. Their primary aim was to clarify ownership or use rights, with a subsidiary aim being to reduce boundary conflicts. The assumption was that the process of demarcating and publicly adjudicating land boundaries followed by registration of landholders' ownership or use rights in a state-administered tenure system would enhance their tenure security. Like many donor driven PDR programs, the program theories for PDR programs in Benin, Ethiopia, and Rwanda were implicitly or explicitly rooted in neoclassical property rights theory about the relationship between tenure security and landholder investment in practices aimed at improving land productivity.

As summarized by Besley (1995), property theorists envision several mechanisms by which secure tenure motivates agricultural investments by landholders and ultimately increases agricultural productivity. First, tenure security fosters investment by assuring the landholder that their land will not be expropriated, and that they will benefit from investments they make in the land. Second, tenure security facilitates land transfers, since prospective renters or buyers have assurance that the person renting out or selling the land has the right to do so. Third, tenure security makes it easier for landholders to use their land as collateral for acquiring capital for investments, since the lender has assurance that the landholder has the right to transfer the land should they default on their debt. Goldstein et al. (2018) note a fourth pathway: tenure security is theorized to reduce the time and cost of defending land rights, thereby freeing resources for productive investment. However, not all mechanisms apply in all contexts. For example, in Ethiopia, where land sales are illegal, one might expect to see an increase in rentals, rather than in land sales. Drawing on evidence from earlier titling programs, current PDR program theories of change typically assume that accompanying measures, such as joint titling for spouses, land rights awareness programs for women, and legal aid that enables less powerful groups to assert land claims, are needed to ensure that women and other marginalized groups can benefit from PDR programs. Note, however, that joint titling in both patrilineal and matrilineal societies is in direct contradiction to the descent group landholding system, and has been resisted by men, as well as sisters and daughters of the men belonging to the descent group in question.

Land redistribution (LRD) programs seek to redress the consequences of poverty and landlessness resulting from concentration of large holdings of farming land in the hands of a comparatively small percentage of the rural population. The African nationalist struggles in South Africa, Zimbabwe, and Namibia in the second half of the 20th century were fueled to a considerable degree by popular grievances over the concentration of good agricultural land in the hands of minority white, European settler communities. African communities had been displaced from land traditionally held under customary tenure, to so-called ethnic homelands, characterized by overcrowding and serving ultimately as reserves of low-wage labor for commercial farms, mines, and urban factories. Classic redistributive land reform theory, derived from the Latin American and Asian experience, holds that breaking up large-scale, plantation style farming units into smaller holdings, distributed to former farm workers and tenant farmers, will result in higher levels of agricultural productivity and higher and better distributed farm-family incomes (Dorner and Kanel 1971), thereby addressing both productivity and social inclusivity concerns. In the Latin American and Asian contexts, title registered in cadastral systems was the dominant form of land registration and the standard of tenure security. Redistributive programs were likely to have greater success when accompanied by a variety of training, marketing, and credit programs tailored to the needs of new, often inexperienced farmers. Importantly, however, land reform in Latin America and Asia was first and foremost a political imperative, a response to colonial legacies of concentration of wealth and power. Social scientists helped make an economic case for land reform, but redistributive economic outcomes served the larger popular political demands of the citizenry advocating a more equitable and just post-colonial order. LRD initiatives are less common in SSA than PDR interventions, but recent LRD programs in South Africa, Namibia, and Zimbabwe were driven by demands for redistributive justice that accompanied the anti-apartheid and independence movements in South Africa and Namibia and the independence struggle in Zimbabwe.

References

Benjaminsen, T. (2002). Formalising land tenure in rural Africa. *Forum for Development Studies*, *29*, 362–366. https://doi.org/10.1080/08039410.2002.9666212.

Besley, T. (1995). Property rights and investment incentives: theory and evidence from Ghana. *Journal of Political Economy*, *103*(5), 903–937. https://doi.org/10.12691/jcd-1-1-2.

Bøås, M. (2009). "New" nationalism and autochthony –tales of origin as political cleavage. *Africa Spectrum*, *44*(1), 19–38. https://doi.org/10.1177/000203970904400103.

Boone, C. (2007). Property and constitutional order: land tenure reform and the future of the African state. *African Affairs*, *106*(425), 557–586. https://doi.org/10.1093/afraf/adm059.

Boone, C. (2015). Land tenure regimes and state structure in rural Africa: implications for forms of resistance to large-scale land acquisitions by outsiders. *Journal of Contemporary African Studies*, *33*(2), 171–190. https://doi.org/10.1080/02589001.2015.1065576.

Boone, C. (2019). Legal empowerment of the poor through property rights reform: tensions and trade-offs of land registration and titling in sub-Saharan Africa. *The Journal of Development Studies*, *55*(3), 384–400. https://doi.org/10.1080/00220388.2018.1451633.

Childress, M., Spievack, D., Varela, D., & Ameyaw, D. (2018, March 20). *Measuring Citizen Perceptions of Tenure Security: Test Surveys of the Global Land Rights Index (Prindex) in Tanzania, Colombia, and India*. Annual World Bank Land and Poverty Conference, Washington, DC. https://www.globallandalliance.org/articles/measuring-citizen-perceptions-of-tenure-security-test-surveys-of-the-global-land-rights-index-prindex-in-tanzania-columbia-and-india.

Chimhowu, A. (2019). The 'new' African customary land tenure. Characteristic, features and policy implications of a new paradigm. *Land Use Policy*, *81*, 897–903. https://doi.org/10.1016/j.landusepol.2018.04.014.

Cotula, L. (2007). Introduction. In L. Cotula (Ed.), *Changes in "Customary" Land Tenure Systems in Africa* (pp. 5–14). IIED and FAO. https://www.semanticscholar.org/paper/Changes-in-%22customary%22-land-tenure-systems-in-Cotula/8b2e9bacef7a14757abc8949f6644765a1f971e6.

Cousins, B. (2005). Tenure reform in South Africa: titling versus social embeddedness. *Forum for Development Studies*, *32*(2), 415–442. https://doi.org/10.1080/08039410.2005.9666322.

Cousins, B. (2007). More than socially embedded: the distinctive character of 'communal tenure' regimes in South Africa and its implications for land policy. *Journal of Agrarian Change*, *7*(3), 281–315. https://doi.org/10.1111/j.1471-0366.2007.00147.x.

Dorner, P., & Kanel, D. (1971). *The Economic Case for Land Reform: Employment, Income Distribution, and Productivity*. Land Tenure Center, University of Wisconsin.

Gezon, L. L. (2002). Marriage, kin, and compensation: a socio-political ecology of gender in Ankarana, Madagascar. *Anthropological Quarterly*, *75*(4), 675–706. https://doi.org/10.1353/anq.2002.0060.

Goldstein, M., Houngbedji, K., Kondylis, F., O'Sullivan, M., & Selod, H. (2018). Formalization without certification? Experimental evidence on property rights and investment. *Journal of Development Economics*, *132*, 57–74. https://doi.org/10.1016/j.jdeveco.2017.12.008.

Honig, L. (2017). Selecting the state or choosing the chief? The political determinants of smallholder land titling. *World Development*, *100*, 94–107. https://doi.org/10.1016/j.worlddev.2017.07.028.

Jayne, T. S., Muyanga, M., Wineman, A., Ghebru, H., Stevens, C., Stickler, M., Chapoto, A., Anseeuw, W., Westhuizen, D., & Nyange, D. (2019). Are medium-scale farms driving agricultural transformation in sub-Saharan Africa? *Agricultural Economics*, *50*(S1), 75–95. https://doi.org/10.1111/agec.12535.

Krantz, L. (2015). *Securing Customary Land Rights in Sub-Saharan Africa Learning from New Approaches to Land Tenure Reform* (Working Papers in Human Geography). Department of Economy and Society. Götesborgs Universitet. https://www.academia.edu/21182243/Securing_Customary_Land_Rights_in_Sub_Saharan_Africa.

Lavigne Delville, P., & Moalic, A.-C. (2019). Territorialities, spatial inequalities and the formalization of land rights in Central Benin. *Africa*, *89*(2), 329–352. https://doi.org/10.1017/S0001972019000111.

Lawry, S., Samii, C., Hall, R., Leopold, A., Hornby, D., & Mtero, F. (2017). The impact of land property rights interventions on investment and agricultural productivity in developing countries: a systematic review. *Journal of Development Effectiveness*, *9*(1), 61–81. https://doi.org/10.1080/19439342.2016.1160947.

Migot-Adholla, S. E., & Bruce, J. W. (1994). Introduction: are indigenous African tenure systems insecure? In J. W. Bruce & S. E. Migot-Adholla (Eds.), *In: Searching for Land Tenure Security in Africa*. World Bank. https://www.google.com/url?sa=t&rct=j&q=&esrc=s&source=web&cd=&ved=2ahUKEwje6sPImIT8AhVgGjQIHSpxD7gQFnoECCwQAQ&url=https%3A%2F%2Fdocuments1.worldbank.org%2Fcurated%2Fen%2F630121468742824113%2Fpdf%2F280431 0paper.pdf&usg=AOvVaw1MJnVtzHhtnMJCysFxJFdI.

Mkodzongi, G. (2016). 'I am a paramount chief, this land belongs to my ancestors': the reconfiguration of rural authority after Zimbabwe's land reforms. *Review of African Political Economy*, *43*(sup1), 99–114. https://doi.org/10.1080/03056244.2015.1085376.

Moyo, S. (2007). Land in the political economy of African development: alternative strategies for reform. *Africa Development*, *32*, 1–34. https://doi.org/10.4314/ad.v32i4.57319.

Mutopo, P. (2011). Women's struggles to access and control land and livelihoods after fast track land reform in Mwenezi District, Zimbabwe. *The Journal of Peasant Studies*, *38*(5), 1021–1046. https://doi.org/10.1080/03066150.2011.635787.

Okoth-Ogendo, H. W. (2002). *The Tragic African Commons: A Century of Expropriation, Suppression and Subversion* (Occasional Paper 24; Land Reform and Agrarian Change in Southern Africa). Programme for Land and Agrarian Studies. University of the Western Cape.

Peters, P. E. (2009). Challenges in land tenure and land reform in Africa: anthropological contributions. *World Development*, *37*(8), 1317–1325. https://doi.org/10.1016/j.worlddev.2008.08.021.

Peters, P. E. (2013). Land appropriation, surplus people and a battle over visions of agrarian futures in Africa. *The Journal of Peasant Studies*, *40*(3), 537–562. https://doi.org/10.1080/03066150.2013.803070.

Peters, P. E. (2019). Revisiting the social bedrock of kinship and descent in the anthropology of Africa. In R. Grinker, S. Lubkemann, C. Steiner, & E. Gonçalves (Eds.), *A Companion to the Anthropology of Africa* (pp. 33–62). https://doi.org/10.1002/9781119251521.ch2.

Peters, P. E. (2020). The significance of descent-based 'customary' land management for land reform and agricultural futures in Africa. In C. M. O. Ochieng (Ed.), *Rethinking Land Reform in Africa New Ideas, Opportunities and Challenges* (pp. 70–83). African Development Bank. https://www.afdb.org/en/initiatives-partnerships/african-natural-resources-centre/publications/rethinking-land-reform-africa-new-ideas-opportunities-and-challenges.

Peters, P. E. (2021). Kinship. In A. H. Akram-Lodhi, K. Dietz, B. Engels, & B. McKay (Eds.), *Handbook of Critical Agrarian Studies* (pp. 139–149). Edward Elgar Publishing. https://www.elgaronline.com/display/edcoll/9781788972451/9781788972451.00024.xml.

Platteau, J.-P. (1996). The evolutionary theory of land rights as applied to sub-Saharan Africa: a critical assessment. *Development and Change*, *27*(1), 29–86. https://doi.org/10.1111/j.1467-7660.1996.tb00578.x.

Tsikata, D. (2016). Gender, land tenure and agrarian production systems in sub-Saharan Africa. *Agrarian South: Journal of Political Economy*, *5*(1), 1–19. https://doi.org/10.1177/2277976016658738.

Turner, M. D., McPeak, J. G., Gillin, K., Kitchell, E., & Kimambo, N. (2016). Reconciling flexibility and tenure security for pastoral resources: the geography of transhumance networks in eastern Senegal. *Human Ecology*, 1–17. https://doi.org/10.1007/s10745-016-9812-2.

World Bank Gender and Innovation Lab (WBGIL). (2019). *Impact Evaluation of Access to Land Project in Benin.* Prepared for the Millennium Challenge Corporation. https://thedocs.worldbank.org/en/doc/537351555943343180-0010022019/original/MCCEvaluationReportIEofAccesstoLandProjectinBeninFINAL.pdf.

3 Realist synthesis methodology

Understanding intervention outcomes in complex contexts

We took a realist synthesis approach to the research question, "What contextual factors shape where, how, and for whom particular tenure interventions result in positive outcomes for tenure security, agricultural investment and productivity, and social inclusion?" Realist synthesis, like other review methods, involves systematically identifying relevant studies and then iteratively extracting and synthesizing data. However, where other types of systematic reviews tend to emphasize the extraction and subsequent analysis of data quantifying program outcomes, realist synthesis is a qualitative research approach that involves the extraction of narrative descriptions of program contexts, mechanisms, and outcomes for the purposes of eliciting a program theory that describes how the program actually works—as distinct from how it is assumed to work—in particular contexts. In short, realist syntheses seek to reveal differences between how interventions are presumed to work (i.e., program theory) and how they work in particular contexts (Jagosh 2019).

Realist synthesis assumes that it is the mechanisms that are triggered by the intervention that result in particular outcomes, with mechanisms being defined as the "underlying entities, processes, or structures which operate in particular contexts to generate outcomes of interest" (Astbury and Leeuw 2010: 368). The mechanisms triggered will vary by the context (Pawson and Tilley 1997). In realist synthesis, the intervention is not the mechanism; rather, it triggers the mechanisms, which consist of the combination of resources provided together with the affected person's response. Realist synthesis seeks to develop context-mechanism-outcome (CMO) configurations that describe intervention outcomes by describing the mechanisms that result in those outcomes and the contextual factors that trigger such mechanisms (Durham and Bains 2015; Pawson and Tilley 1997). In our analysis, for each case, we developed an initial CMO configuration. Because of the complexity of the interventions, all the cases had multi-strand CMOs. Once we had explored the intervention and the resulting mechanism and its outcomes and associated contextual factors, we modified the CMOs for each case to reflect how the intervention had worked in practice.

DOI: 10.4324/9781003365679-3

Our inclusion/exclusion criteria are outlined in Table 3.1. We limited our search to articles published post-2010 to increase the likelihood that sufficient time would have passed for outcomes to be discernible. We limited our systematic search to peer review journal articles and impact evaluations available through the International Initiative for Impact Evaluation (3iE) and Web of Science (WOS). Search terms were drawn from the land tenure and land governance literature (Lisher 2019; Robinson et al. 2018; Scalise and Giovarelli 2020) and our outcomes of interest (See Annex 1: Search terms).

Through a preliminary review of the initial search results by title, abstract, and text scan, we identified 86 journal articles and 14 impact evaluations as potential candidates for inclusion in our analysis. We then conducted purposive searches to fill in gaps from our systematic searches using Google Scholar and our own libraries. The purposive search yielded 50 additional papers, 13 of which met our inclusion criteria for further appraisal. Following realist synthesis methodology, we later drew on literature on theoretical frameworks, legal documents, and articles regarding history and social systems for the case study countries as background for enriching our understanding of contextual factors that might have impacted the interventions.

Using a customized appraisal form, we appraised the candidate sources to ascertain their relevance, quantity, and quality of information on intervention, outcomes, and context; and methodological quality (see Annex 2). The appraisal form is in Annex 3.

We focused on agricultural productivity and social inclusion as our final outcomes of interest. We have seen that theoretical assumptions about weak security under customary tenure have been central features of land reform inventions in SSA designed to increase agricultural productivity through formal rights recognition. In the past two decades, designers of tenure interventions have sought to ensure that marginalized groups, such as women, youth, and migrants also benefit from these interventions. We therefore included social inclusivity as a second outcome of interest. We included tenure security and investment as intermediate outcomes of interest, so as not to assume that the intervention had provided tenure security or incentivized investments that would result in increased agricultural productivity or social inclusion.

Through this appraisal process, we identified 33 articles for inclusion in the analysis. Additional purposive searches along with a review of included article bibliographies yielded an additional ten articles, for a total of 43 studies drawn from 14 countries (Benin, Burkina Faso, Ethiopia, Ghana, Ivory Coast, Kenya, Liberia, Malawi, South Africa, Rwanda, Tanzania, Uganda, Zambia, Zimbabwe). We decided that at least three different sources were needed for a country to have sufficient coverage to be included in the analysis. Only four countries met this criterion: Benin, Ethiopia, Rwanda, and Zimbabwe. Once the countries of interest were identified, we conducted additional searches

Table 3.1 Inclusion and exclusion criteria (all databases) (for a description of the quality appraisal protocol used to assess quality, see Annex 3)

Criteria	*Inclusion*	*Exclusion*
Publication type	Peer-reviewed publications, published working papers, published organizational research reports on interventions	Literature on theoretical frameworks, legal documents, documents that do not review a specific intervention
Publication date	January 1, 2010, to May 30, 2020	2009 and earlier
Language	English, French, Portuguese	Language other than English, French, or Portuguese
Study type	Project evaluations, econometric studies, ethnographic studies, mixed-method studies	Theoretical studies, essays, frameworks
Geography	Low- and middle-income countries in SSA[a]	Countries outside SSA
Land use type	Rural areas	Urban; forest; grazing land
Intervention type	Land tenure security initiatives (including for women), for individual or collectively held lands (e.g., parcel demarcation/ registration, land redistribution programs, state recognition of customary lands)	Land interventions without a tenure focus (e.g., agricultural extension programs) Legal aid and legal education programs Alternative dispute resolution programs
Outcome types	Investment, productivity, social equity,[b] resilience[c]	Other outcomes (e.g., climate change)

[a] Including: Angola, Benin, Botswana, Burkina Faso, Burundi, Cameroon, "Cape," Central African Republic, Chad, Comoros, Congo, Brazzaville, Côte d'Ivoire, Djibouti ,Guinea Eritrea, Ethiopia, Gabon, Gambia, Ghana, Kenya, Lesotho, Liberia, Madagascar, Malawi, Mali, Mauritania, Mauritius, Mozambique, Namibia, Niger, Nigeria, Réunion, Rwanda, Sao Tome, Senegal, Seychelles, Sierra Leone, Somalia, South Africa, Sudan, Swaziland, Tanzania, Togo, Uganda, Western Sahara Zambia, Zimbabwe.

[b] Land markets were included under social equity.

[c] Resiliency was subsequently dropped as an outcome of interest.

to locate supplemental materials to provide a richer picture of the context in which the interventions took place.

For data extraction, we used a customized form to record contextual factors, mechanisms, and outcomes of interest for each study (see Annex 4). We used a grounded theory approach to identifying the contextual factors, mechanisms and outcomes, drawing on the evidence presented in each study. We extracted data about how the studies defined the problem to be solved by the intervention, the intervention's design features, the theorized pathway toward our outcomes of interest, and the results for our outcomes of interest. We also extracted data about the existing tenure system, including the rights of women and other vulnerable groups, as well as household production orientation. We then identified cross-cutting themes across the studies, relative to outcomes of interest and contextual factors. Next, we developed charts of results (positive, neutral, or negative/no impact) by country for each set of outcomes (see Annex 5).

The in-depth analysis of these four countries consisted of a multi-step, iterative thematic analysis process. We adopt a pragmatic stance to research using abductive reasoning, moving between deduction and induction (Kaushik and Walsh 2019). We translated the program theories that were explicitly articulated in the studies identified through our search into CMO configurations. We then reviewed the data extraction sheets and articles to identify the contextual factors associated with specific outcomes. At this point we identified additional articles for Ethiopia (1), Rwanda (3), and Zimbabwe (1) to include in the in-depth analysis. Ultimately, our in-depth analysis centered on 29 studies. Following realist synthesis practice, we selected the countries for inclusion in the analysis based on the amount of evidence rather than representativity. By happenstance, the countries with sufficient evidence were drawn from West Africa, Central Africa, East Africa, and Southern Africa. Our study thus includes a different set of countries than those most frequently covered in the English language literature on land tenure in SSA (i.e., Kenya, Ghana, Nigeria).

Once the data was extracted for the CMOs, we constructed narratives describing the tenure intervention, mechanisms, outcomes, and key contextual factors for each country. The CMO structures in all four countries incorporate tenure security, investment, agricultural productivity, and social inclusion. For each country, we summarize the literature that evaluated the impacts of tenure interventions on tenure security, investment in land improvements, agricultural

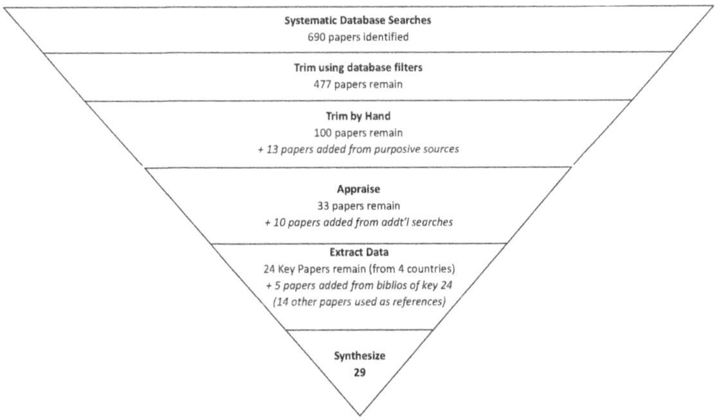

Figure 3.1 Realist synthesis literature identification process.

productivity, and social inclusion. We then draw on these evaluations and supplementary materials that provide contextual data for the case study countries to identify key contextual factors that we hypothesize may have affected the outcomes identified in the evaluation literature. In the discussion section, we compare findings across the countries to consider how country-level context differences may have affected PDR outcomes more generally. The entire team reviewed the narratives to identify common patterns as well as unique responses and corresponding contexts. Figure 3.1 depicts the stages involved in identifying relevant literature, along with the number of papers included after each stage. In the final stage of the analysis, we revised the PDR and LRD CMOs in light of our findings for each of the case study countries.

References

Astbury, B., & Leeuw, F. L. (2010). Unpacking black boxes: mechanisms and theory building in evaluation. *American Journal of Evaluation, 31*(3), 363–381. https://doi.org/10.1177/1098214010371972.

Durham, J., & Bains, A. (2015). Research protocol: a realist synthesis of contestability in community-based mental health markets. *Systematic Reviews, 4*(1), 32. https://doi.org/10.1186/s13643-015-0025-3.

Jagosh, J. (2019). Realist synthesis for public health: building an ontologically deep understanding of how programs work, for whom, and in which contexts. *Annual Review of Public Health, 40*(1), 361–372. https://doi.org/10.1146/annurev-publhealth-031816-044451.

Kaushik, V., & Walsh, C. A. (2019). Pragmatism as a research paradigm and its implications for social work research. *Social Sciences, 8*(9), 255. https://doi.org/10.3390/socsci8090255.

Lisher, J. (2019). *Guidelines for Impact Evaluation of Land Tenure and Governance Interventions*. International Fund for Agricultural Development and Global Land Tool Network. https://www.ifad.org/en/web/knowledge/-/publication/the-guidelines-for-impact-evaluation-of-land-tenure-and-governance-interventions.

Pawson, R., & Tilley, N. (1997). *Realistic Evaluation*. Sage.

Robinson, B. E., Masuda, Y. J., Kelly, A., Holland, M. B., Bedford, C., Childress, M., Fletschner, D., Game, E. T., Ginsburg, C., Hilhorst, T., Lawry, S., Miteva, D. A., Musengezi, J., Naughton-Treves, L., Nolte, C., Sunderlin, W. D., & Veit, P. (2018). Incorporating land tenure security into conservation: conservation and land tenure security. *Conservation Letters, 11*(2), e12383. https://doi.org/10.1111/conl.12383.

Scalise, E., & Giovarelli, R. (2020). *What Works for Women's Land and Property Rights? What We Know and What We Need to Know*. Research Consortium on Women's Land Rights and Resource Equity.

4 Benin's Plan Foncier Rural program (rural land plan program)

Political economy context

Benin's land tenure system is an amalgam of indigenous institutions that existed prior to the establishment of French colonial rule in 1894, colonial institutions, and institutions that have emerged since Benin gained independence in 1960. When the French took control over the area that is now known as Benin, they encountered a mix of governance systems, ranging from the large, hierarchical and centralized Dahomey kingdom to numerous smaller and non-hierarchical political entities in frontier areas (Platteau 2019). The French initially ruled indirectly through a series of protectorates, but in the early 20th century, they introduced a more direct form of administration that fundamentally transformed social relations between communities (Le Meur 2006). The colonial regime divided the country into political units that were demarcated and distributed in a way that served its interests of maximizing exports of raw materials while maintaining social order (Boone 2007). Indigenous chiefs who were viewed as allies were selected to rule at the canton and village level (Le Meur 2006). The administrative village was a French territorial creation, and did not necessarily reflect existing socio-political units, laying a foundation for inter-village conflict (Lavigne Delville and Moalic 2019). Nor were the individuals who were designated by the colonial regime as village chiefs necessarily viewed by the local population as having legitimate authority to allocate land and adjudicate land conflicts (Le Meur 2006). The French introduced the concept of state domainality, in which all land not registered with the state as private land was considered state land (Lavigne Delville 2019). At independence, Benin opted to retain the colonial administrative structure inherited from the French as well as much of the colonial legal framework, including laws governing the allocation and adjudication of land (Lavigne Delville 2019). More recently, Lavigne Delville et al.'s (2017) study of land markets in West Africa, including Benin, found that there is strong demand for land from both external and internal investors. In Benin, demand for land is strongest in peri-urban areas around major cities and in areas that are close to transportation infrastructure and markets (Lavigne Delville 2019).

DOI: 10.4324/9781003365679-4

A notable feature of Benin's customary tenure system is the tutorat system which Chauveau (2006) famously described for Ivory Coast, and which is also widespread in central Benin's frontier areas where labor is scarce and arable land still abundant (Lavigne Delville and Moalic 2019). In the tutorat system, members of a village's founding lineage grant farming rights to newcomers. These rights are permanent and transmissible, as well as deeply embedded in ongoing social relationships between the host and the guest (Lavigne Delville and Moalic 2019: 4). The tutorat system also operates at the collective scale. Founding villagers may grant migrant families the right to establish hamlets or satellite villages, a practice that allows the founders to expand and retain territorial control (Lavigne Delville and Moalic 2019). A hierarchy is thus created within the traditional village system, with some villages dependent on others for their land rights. Adding to this complexity is the state's system of administrative villages, which is superimposed over the traditional village system and is structured such that it is not uncommon for administrative villages to encompass all or parts of territories belonging to multiple traditional villages within their boundaries (Lavigne Delville and Moalic 2019). Over the years, administrative villages have acquired more political power than other villages, including those that were traditionally more powerful. In cases where a satellite village is designated as an administrative village, the power balance between founding and satellite village may shift in favor of the satellite village (Lavigne Delville and Moalic 2019).

Benin became a multi-party democracy in 1989–1990 (Bierschenk and de Sardan 2003), and until recently was considered to be among the most politically stable countries in Africa. Bierschenk and de Sardan (2003) describe Benin's local level politics as very fragmented, with a weak centralized state and many different authorities having local governance legitimacy. Decentralization efforts began in late 1980s and early 1990s but have been slow to make headway due in part to resistance from local-level traditional authorities and in part to weak political will and capacity on the part of the state (Bierschenk and de Sardan 2003). Corruption is widespread, including in the land sector (Lavigne Delville 2019). Customary governance systems in Benin are not the systems that existed in the pre-colonial period, but rather are locally based systems that have adapted to changing social, political, and economic conditions (Bierschenk and de Sardan 2003). The key point is that they operate in parallel to the state and are viewed as legitimate by the people in the areas where they are operational (Bierschenk and de Sardan 2003).

Program overview

The tenure intervention examined in Benin was the Millennium Challenge Account's (MCA) Plan Foncier Rural (PFR—Rural Land Plan) program (hereafter referred to as PFR), which was implemented between 2006 and 2011.

The PFR program was a village-wide approach designed to identify, map, and register individual and household rights within customary tenure regimes (Goldstein et al. 2018). Adapted from a similar approach initiated in Ivory Coast in 1989, the program sought to incentivize land improvements that would increase Investments in agricultural productivity; it also aimed to reduce land conflicts and catalyze a more robust land market (WBGIL 2019). PFRs were first piloted in several regions of Benin during the 1990s with funding from European donor agencies (WBGIL 2019) and institutionalized through the Rural Land Tenure Act 2007 (Loi portant régime foncier rural), which established a new legal category for rural landholdings, the Certificat Foncier Rural (CFR, or Rural Land Certificate) (Lavigne Delville 2019). The 2007 Act legalized plots registered in the PFRs, and established a new land administration framework, grounded in rural communities (Lavigne Delville 2019). The MCA provided funding for Benin to scale the PFR program to the national level (Goldstein et al. 2018).

The core elements of the program included establishing village-level land management commissions, conducting socio-legal studies to identify owners and users of parcels and validate land claims, surveying and demarcation of parcel boundaries, and development of village land use plans (Goldstein et al. 2018). A village-wide approach to parcel demarcation and rights recording was adopted, with the goal of including all of the parcels within each village's boundaries in the final village land plan. The program was designed to permit the recording of primary individual and collective rights and rights holders, as well as secondary (delegated) rights and rights holders during the mapping process (Giovarelli et al. 2015).

Once the village land use plan was completed, a landholder whose rights had been recorded could apply to the land commission for a land certificate. The land commission then forwarded the request to the commune for review and approval (WBGIL 2019). The program supported the establishment of land commissions at the commune level, as well as training of land commission staff, who were responsible for delivery of land certificates. However, the project did not directly support certificate delivery (WBGIL 2019). Goldstein et al. (2015: 8) describe the right provided by a land certificate as similar to that provided by a title. At the time the PFR program was implemented, the law provided a pathway by which a land certificate could be converted into an ownership title (Lavigne Delville 2019).

In roughly one-quarter of the communes, the project hired a consulting firm to identify secondary rights holders, ascertain their landholding needs, and develop social programs to address those needs (Giovarelli et al. 2015). Examples of such programs included: social contracts, common resources management plans, and systematic recording of tenancy and land use contracts. Additionally, the PFR program guidelines recommended that joint use rights certificates be issued for married couples (Giovarelli et al. 2015).

The program was carried out in 40 communes scattered throughout the country, with the 300 participating villages selected through a lottery. To be eligible for participation in the project, villages had to meet certain criteria, including having significant levels of agricultural production, high levels of poverty, general acceptance for women's rights, particularly with respect to land inheritance, and presence of land conflicts and disputes (WBGIL 2019).

When the program ended in 2011, PFRs had been completed in 294 villages, more than 72,000 parcels had been mapped, and roughly 68,700 landholders had had their rights recorded in village PFRs (WBGIL 2019). However, at the time of the final evaluation (four years post-completion), rural land certificates had been issued for only 19% of the parcels demarcated through the MCA's PFR program and only 8% of households in PFR villages had obtained land certificates (WBGIL 2019).

Program theory for Benin's PFR program

Goldstein et al. (2018) describe the goals of the PFR program to be increasing tenure security through the demarcation and mapping of parcels, and through the opportunity to obtain certificates from the state that attest to who owns the parcel. Table 4.1 depicts the program theory described by Goldstein et al. (2018), including anticipated outcomes with respect to agricultural productivity, land transactions, and access to credit. The PFR program had two other major desired outcomes: ensuring fair and equitable access to land for women and other secondary rights holders (WBGIL 2019) and reducing land conflicts between individuals (Lavigne Delville and Moalic 2019; Yemadje et al. 2014) and between villages (Wren-Lewis et al. 2020). The theorized pathways for these outcomes are depicted in Table 4.2.

As indicated in Tables 4.1 and 4.2, the MCA's PFR program in Benin consisted of multiple non-exclusive strands. The top row of both tables lists contextual characteristics identified in the literature as the reasons for implementing a land registration process. Underneath the contextual factors, we list the different pathways that are theorized in the PFR program theory. For each theorized pathway, the subsequent rows describe its theorized mechanism, which consists of a resource and a response, followed by intermediate outcomes and the final outcome.

Realized outcomes

The presence of land conflicts was the only measure of tenure security common to all of the Benin studies that were included in the synthesis. Other measures of tenure security included the presence of clear boundaries (Goldstein et al. 2018; WBGIL 2019; Wren-Lewis et al. 2020), levels of trust in the ability of local land institutions to resolve disputes (Wren-Lewis et al. 2020), perceptions of rights to sell a parcel (Goldstein et al. 2015), whether landholders are willing to risk

Table 4.1 Theoretical CMO for investment, land transactions, and access to credit for Benin's Plan Foncier Rural program

1. Key tenure-related contextual factors identified by program designers as influencing agricultural productivity

- Presence of overlapping customary and state tenure systems, which was assumed to be an important source of insecurity.
- Costly and time-consuming procedures for registering land rights through the state land administration system which discouraged rural smallholders from seeking state-issued land titles (Goldstein et al. 2018; WBGIL 2019; Wren-Lewis 2020).
- Presence of land conflicts, which project designers attributed in part to the lack of state-issued documentation of land rights and absence of locally accessible land registries (WBGIL 2019; Wren-Lewis 2020; Yemadje et al. 2014).
- Insecure customary tenure associated with lack of rights documentation was identified as an important disincentive for rural households to engage in land practices that increase agricultural productivity (WBGIL 2019; Yemadje et al. 2014).
- In rural areas, women tend to have derived access and use rights to land and other resources, and thus their tenure is thought to be relatively insecure (Giovarelli et al. 2015).

2a. Theorized pathway following PFR implementation for achieving investment in land management practices	**2b. Theorized pathway following PFR implementation for facilitating of land transactions**	**2c. Theorized pathway following PFR implementation for improved access to credit**
3a. Mechanism *Resource*: clear and enforceable land rights for landholders and users *Response*: improved perceived tenure security provides confidence for holders that their farm yields won't be disputed	**3b. Mechanism** *Resource*: clear and enforceable land rights for landholders *Response*: potential buyer or renter has assurance that the landholder has the right to transfer ownership or use rights, facilitating the sale and transfer of land to most productive farmers	**3c. Mechanism** *Resource*: clear and enforceable land rights for landholders *Response*: potential lender has assurance that the landholder has the right to transfer ownership rights; this enhances the collateral value of land, leading to easier access to credit for landholders and potential buyers
4a. Intermediate outcome 1 Greater investment in land management practices with long-term benefits (i.e., tree-planting, planting of perennial crops, construction of anti-erosion terraces, long-term fallows)	**4b-1. Intermediate outcome 1** Landholders who lack the capacity to farm the land can easily rent it out or sell it to someone who is able to farm it **4b-2. Intermediate outcome 2** Purchaser or renter with more assets can put the land into more efficient production	**4c-1. Intermediate outcome 1** Improved land market functioning (rental and sales) as third parties such as mortgage lenders easily access reliable land and ownership information **4c-2. Intermediate outcome 2** Landholders use borrowed funds to make investments in their land that improve soil fertility (i.e., application of improved fertilizers) or reduce erosion (i.e., tree-planting; construction of anti-erosion structures)

5. Final outcome: agricultural productivity increases

Table 4.2 Theoretical CMOs for reduced land conflict and social inclusion for Benin's Plan Foncier Rural program

<table>
<tr><td colspan="4">1. Key tenure-related contextual factors identified by program designers as influencing land conflicts and social inclusion
• Presence of overlapping customary and state tenure systems, which was assumed to be an important source of insecurity
• Costly and time-consuming procedures for registering land rights through the state land administration system which discouraged rural smallholders from seeking state-issued land titles (Goldstein et al. 2018; WBGIL 2019; Wren-Lewis 2020)
• Presence of land conflicts, which project designers attributed in part to the lack of state-issued documentation of land rights and absence of locally accessible land registries (WBGIL 2019; Wren-Lewis 2020; Yemadje et al. 2014)
• Insecure customary tenure associated with lack of rights documentation was identified as an important disincentive for rural households, including secondary rights holders, to engage in land practices that increase agricultural productivity (WBGIL 2019; Yemadje et al. 2014)
• In rural areas, women tend to have derived access and use rights to land and other resources, and thus their tenure is thought to be relatively insecure (Giovarelli et al. 2015)</td></tr>
<tr><td colspan="2">2a. Theorized pathways to fewer conflicts following PFR implementation</td><td colspan="2">2b. Theorized pathways for social inclusion following PFR implementation</td></tr>
<tr><td>3a. Farmer-to-farmer conflict</td><td>3b. Inter-village conflicts</td><td>3c. Access to land for women</td><td>3d. Access to land for secondary or overlapping rights holders (youth, borrowers, tenants, sharecroppers)</td></tr>
<tr><td>4a. Mechanism
Resource: clear and enforceable land rights for landholders and users
Response: potential rival land claimants believe their claim will not be upheld should they try to expropriate the land or encroach on boundaries; hence they have an incentive to respect borders</td><td>4b. Mechanism
Resource: clear and enforceable rights for villages
Response: neighboring villages recognize that efforts to claim land from other villages are unlikely to be upheld</td><td>4c. Mechanism
Resource: clear and enforceable land rights for secondary rights holders (women and children) created through including names of female spouses on land certificates and targeting women in female-headed households for certification
Response: women feel more secure, providing confidence that if they make investments in land that they or their heirs will benefit</td><td>4d. Mechanism
Resource: clear and enforceable land rights for secondary rights holders created through inclusion of names of all rights holders on land certificates, and, in Benin, through written locally recorded witnessed records
Response: improvement in perceived tenure security provides confidence for holders that their farm yields won't be disputed</td></tr>
</table>

5a. Intermediate outcome: Fewer conflicts over land among farmers; less time devoted to guarding property and engaging in conflict, more time in productive activity including agriculture	**5b. Intermediate outcome:** Fewer conflicts between villages over land: less time devoted to guarding property and engaging in conflict, more time in productive activity including agriculture	**5c. Intermediate outcomes:** Reduced eviction threats for married women upon death or divorce Equitable access to land for daughters and sons upon succession and inheritance Women participate in land use decisions	**5d. Intermediate outcome:** Secondary rights holders invest in soil enhancement/conservation practices such as fallowing, anti-erosion terraces, tree-planting, etc.
6a. Final outcome: Agricultural productivity increases	**6b. Final outcome:** Agricultural productivity increases	**6c. Final outcome:** Women benefit from profits derived from land use and transferability of land; increased agricultural productivity	**6d. Final outcome:** Agricultural productivity increases, and secondary rights holder's wellbeing improves

leaving their land fallow (Goldstein et al. 2015, 2018), formal documentation of land rights (Goldstein et al. 2018; WBGIL 2019; Yemadje et al. 2014), written documentation other than a title or certificate (Yemadje et al. 2014), and witnessed but undocumented tenure arrangements (Yemadje et al. 2014).

Land conflicts

Using land conflicts as an indicator for tenure security, PFR's tenure security outcomes were mixed (see Annex 5 for additional details on outcomes). Goldstein et al. (2018) and WBGIL (2019) found no impact on the incidence of land conflicts; Wren-Lewis et al. (2020) reported a small reduction in intra-village land conflicts and a somewhat larger reduction in inter-village conflicts. However, they note that it is unclear whether the observed positive outcomes were due to the boundary marking process, resolution of land disputes through the demarcation process, formation and operation of the land committees, or a combination of these program components.

On the Adja Plateau, which is located in a densely populated area of Southern Benin where land is in high demand and long-term tenancy arrangements are common, PFR exacerbated tensions between tenants and landowners, with tenants reporting being evicted after the program was initiated (Yemadje et al. 2012). To address these conflicts, the commune developed—and the project subsequently adopted—a set of tenancy agreement templates that could be used by landlords and tenants when entering into new agreements.

In the less-densely populated Collines Department, a frontier zone north of the Adja Plateau, efforts to establish PFRs catalyzed several types of conflicts (Lavigne Delville and Moalic 2019). In some cases, conflicts erupted between long-term migrants with access to land through permanent borrowing agreements with founding lineage members over whose name parcels should be registered under. In other cases, conflicts arose between founding village authorities and satellite hamlet authorities, with the latter seeking to wrest territorial control from the former. Lavigne Delville and Moalic (2019) concluded that migrants were particularly likely to lose rights to parcels that they had been farming if the land had been left fallow. They also found that conflicts were much less likely to involve short-term migrants who, as measured by customary norms, had not cultivated borrowed parcels long enough to acquire permanent use and intergenerational transmission rights.

Demand for land certificates

Interest in obtaining land certificates appeared to be limited—only 8% of the households included in the final evaluation had obtained a CFR four years after the project ended (WBGIL 2019). Land commission members and village advisors were more likely to obtain CFRs than other villagers, as were larger households and households whose heads had a higher level of education

(WBGIL 2019). The final evaluation did not provide data from participating villagers as to why they did not obtain CFRs.

Rental activity

In the Adja Plateau region, Yemadje et al. (2012, 2014) reported an increase in plots rented out in PFR villages, which they attribute to landlords' having a greater sense of tenure security. However, their study did not include control villages, so it is unclear whether the differences were caused by the intervention. In contrast, Goldstein et al. (2015) found that land rentals and sharecropping declined in PFR villages. They speculate that landowners may have reclaimed parcels to assert their rights before land certificates were issued, an interpretation that is supported by Yemadje et al.'s (2014) reports of tenants in the Plateau Region being evicted prior to the start of land registration activities.

Yemadje et al. (2014) found that the proportion of paper-based tenancy contracts was higher in the PFR case study village, and more plots were rented out in that village. Yemadje et al. (2014) attribute the higher rate of rentals to the expanding use of written rental contracts, and landowners' subsequent greater level of confidence that tenants will adhere to the rental terms. Subsequent to the PFR program, Yemadje et al. (2014) documented a shift from the use of oral to written contracts and from unwitnessed to witnessed contracts. The long-term impacts of this shift are unknown as they had only recently been introduced at the time of the study.

Investment outcomes

Investment outcomes were generally positive. WBGIL (2019) reported an increase in tree-planting and perennial crops by PFR villagers. In the densely populated Adja Plateau, Yemadje et al. (2014) found that fertilizer use increased among PFR villagers, but oil palm fallowing, a traditional soil fertility improvement practice, had declined. Since the study did not have a control group, exogenous factors such as population growth cannot be ruled out as alternative explanations. The most significant positive investment outcome was that female heads of households in PFR villages were more likely to leave their demarcated (but certificate-lacking) fields in fallow, an outcome that Goldstein et al. (2018) and WBGIL (2019) attribute to a sense of increased security for those parcels, which allowed those women to farm less secure parcels outside the village boundaries.

Agricultural productivity outcomes

Only the interim evaluation (Goldstein et al. 2018) and final evaluation (WBGIL 2019) provided agricultural productivity outcome data. These evaluations showed no impacts overall on agricultural productivity, and a decline in yields for female heads of household.

Women's, tenants', and migrants' outcomes

The interim evaluation showed that awareness of the PFR program and participation in the decision-making processes associated with parcel demarcation and rights recording differed along gender lines (Goldstein et al. 2015), with female heads of households less likely to participate in PFR meetings, be involved in land management commissions, or be aware of the program. However, this did not seem to affect female heads of households' participation in parcel demarcation or their tenure security since male and female heads of households were equally likely to have clearly demarcated borders and a similar likelihood (in both cases very low) of having experienced a land conflict in the previous 12 months (WBGIL 2019). Giovarelli et al.'s (2015) post-project gender assessment found that tenure security may have been weakened for women who were not heads of households, and whose rights were less likely to be registered. In the Adja Plateau, some tenants reported that they were evicted when the land registration project was announced; however, overall rentals increased (Yemadje et al. 2012, 2014). Tenure security for migrants, especially more recent newcomers, was weakened in the Collines Department, a frontier region (Lavigne Delville and Moalic 2019).

Contextual factors affecting outcomes

A combination of widespread legitimacy of traditional authorities in Benin, coupled with reforms that strengthened traditional authority over land and the misfit between a tenure system characterized by nested and overlapping rights and one that simplifies rights, likely resulted in the mixed outcomes observed in Benin for the MCA-funded PFR program. The following section summarizes the key contextual factors we identified as contributing to these outcomes (see Annex 6 for additional details on contextual factors).

Reforms that strengthened customary land governance systems

An important contextual factor likely contributing to the observed limited interest on the part of villagers in obtaining land certificates through the PFR program was the presence of largely functional customary land governance systems in much of rural Benin. These systems were strengthened at the outset of the PFR program when the 2007 Rural Land Tenure Act replaced the presumption of state ownership for untitled land, with the presumption of customary ownership for untitled lands in productive use. Moreover, any benefits attached to having a rural land certificate diminished further when the 2013 Land and Domain Code replaced the rural land certificate with the Attestation de Détention Coutumière (Attestation of Customary Possession) (Lavigne Delville 2019).

Nested and overlapping rights

As in much of SSA, customary landholding systems in Benin are based on the descent group and consist of complex nested systems of overlapping rights to land and natural resources. The PFR process simplified the complex web of rights to demarcated parcels in a way that favored the recording of individual or households'[1] rights rather than descent group rights (Lavigne Delville and Moalic 2019), creating a context ripe for conflict. Migrants, women, and renters were among the categories of land users most at risk of having their tenure security weakened as a result of the PFR process.

First settlers vs. migrants, founding villages vs. satellite villages

Lavigne Delville and Moalic (2019) identify the tutorat or guardianship system of land allocation that prevails in frontier areas[2] in Benin as a factor that explains the limited, and in some cases, negative outcomes of PFR. Lavigne Delville and Moalic (2019) contend that PFR implementation catalyzed land conflict in part because the program was based on a misconception that villages were autonomous governance entities with each having its own territory. They found that PFR-related land conflicts emerged in Collines around parcels farmed by migrants who had use rights to lineage lands. PFR field teams generally registered such parcels in the name of founding lineage members who had allocated the land rather than in the names of the migrants farming the land, particularly when fields had been left in fallow. Migrants who managed to be registered as primary rights holders to lands acquired through the tutorat system, tended to be individuals who had been in the area for a long time, who had planted trees many years previously, or who were politically well-connected (Lavigne Delville and Moalic 2019).

PFR also sparked village-level conflicts over territories occupied by satellite villages or hamlets, with founding villages tending to win out (Lavigne Delville and Moalic 2019). However, satellite villages that had acquired administrative village status under the state system, and which therefore wielded greater political power than their host villages, were better able to ensure that their residents were registered as primary rights holders of lands they farmed (Lavigne Delville and Moalic 2019).

Renters

The social status and political power of different categories of landholders also affected PFR's outcomes in the Adja Plateau, where the project exacerbated tensions between long-term tenants and landowners. Some tenants were evicted after the MCA PFR project was initiated (Yemadje et al. 2012), while others ended up with less favorable terms of access after their landlord registered the land (Yemadje et al. 2014).

State and customary law: women's equity and access

The degree to which state and customary law support equity in women's access to land emerged as important contextual factors affecting women's tenure security. Benin's Constitution recognizes gender equality, and its Persons and Family Code states that wives can inherit land from their husbands. However, wives can only inherit if they are in marriages that are registered with the state. Since 2004, the state no longer recognizes polygamous marriages, so only women in monogamous marriages enjoy these inheritance rights (Giovarelli et al. 2015). The number of women affected is substantial since roughly 41% of women in Benin are in polygamous marriages (INSAE and ICF 2019).

The PFR project encouraged but did not require joint certification for husbands and wives, and, in practice, it proved an elusive goal. Giovarelli et al. (2015) point to the failure of the project implementers to account for the weak bargaining position of secondary rights holders, which meant that women's land rights frequently were not recorded. Women, most of whom in Benin gain access to land through their husband or another male relative, were reluctant to ask to be included as rights holders due to the potential for social friction (Giovarelli et al. 2015). Giovarelli et al. (2015) attribute the program's inability to adequately incorporate secondary rights in the rights recording process to the following contextual factors: (1) poor understanding by policymakers and program teams in how to identify and record secondary rights (2) resistance by primary rights holders, who feared losing control over their land, (3) reluctance on the part of many women to assert their rights for fear of creating social tensions within their households, (4) a project schedule that allocated insufficient time for recording multiple rights, and (5) insufficient attention paid to programming that could strengthen women's bargaining position, such as workshops on women's land rights for village and community leaders. In some communes, the program initiated supplementary social programs, such as training workshops for field teams, and village and community leaders and general public awareness campaigns on equality of access to land and women's land rights. Although the social programs experienced some successes, particularly in providing women with extension services and enabling them to gain access to land through written sharecropping and other tenancy arrangements, the programs were begun too late in the project and applied in too few communes to have widespread impact (Giovarelli et al. 2015).

Revised CMO for Benin: an illustrative example

Our realist synthesis of Benin's PFR program suggests that the neoclassical property rights theory linking land rights and tenure security and investment that underlies it does not adequately account for the complexities of Benin's land tenure and social systems. The PFR program designers were aware of the complexities of customary tenure systems in Benin and tried to build in mechanisms for identifying and recording multiple rights. However, implementation of these mechanisms was flawed, in large part because it is difficult to translate complex rights into a system that is structured for recording simplified rights (Lavigne Delville and Moalic 2019).

Figure 4.1 depicts an alternative CMO for Benin's PFR activity that seeks to capture some of those complexities. A circular design was adopted to emphasize that it is a configuration of contextual factors that interact with tenure interventions to produce mechanisms that lead to outcomes, and to emphasize that CMO configurations need not be linear. This variant of the CMO illustrates how mechanisms and outcomes may differ for male and female-headed households given the contextual factors present in Benin.

The center circle lists the key components of Benin's PFR program, while the outer circle lists factors that influenced the outcomes of interest in the three PDR cases in our study. Block arrows link the two circles. The solid block arrows represent factors that played an important role in shaping Benin's PFR program outcomes. Block arrows that are not filled in indicate important factors that shaped outcomes in Ethiopia and Rwanda, but which were less relevant in Benin. The arrow labeled with a 1 indicates a male head of household who resides in the program area; the arrow labeled with a 2 indicates a female head of household in the same community. The top portion of the box on the right describes the mechanism (i.e., resource and response) triggered by the PFR program and the subsequent intermediate and final outcomes for male heads of households. The bottom portion describes the mechanism triggered by the PFR program and subsequent intermediate and final outcomes for female heads of households.

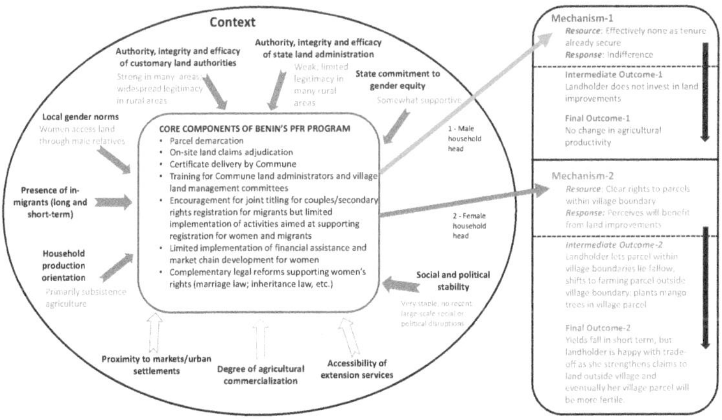

Figure 4.1 Illustrative revised CMO for Benin's PFR program explaining differential outcomes for gender.

In this example, a male head of household who is not a migrant has secure tenure under the prevailing customary tenure system. The PFR program does not provide him with a resource that he lacks, and consequently, does not affect his behavior. A female head of household, in contrast, has relatively insecure tenure under the prevailing system. The PFR program provides her with stronger land rights through the public boundary demarcation and rights recording process and, if she chooses to apply for it, a state-issued land certificate. The female head now feels that she no longer risks losing her demarcated parcel if she doesn't cultivate it for a few years. She therefore lets her demarcated parcel lie fallow so it will regain its fertility and focuses her agricultural activities on a parcel located outside the village boundaries which was not included in the PFR process, and which she feels is less secure.

The illustrative CMO can be adapted to other contexts within Benin. For example, the mechanisms triggered by a land certification program for male heads of household would likely differ in areas close to urban areas where demand for land is high and customary authorities wield less authority.

Notes

1 Collective rights could be registered, but in practice it proved difficult to do so (Lavigne Delville 2019).

2 Lavigne Delville (2019: 11) defines pioneer fronts as "areas where land is still available and where they can settle and create a farm." Pioneer fronts are now mostly located in the center of Benin.

References

Bierschenk, T., & de Sardan, J.-P. O. (2003). Powers in the village: rural Benin between democratisation and decentralisation. *Africa*, *73*(2), 145–173. https://doi.org/10.3366/afr.2003.73.2.145.

Boone, C. (2007). Property and constitutional order: land tenure reform and the future of the African state. *African Affairs*, *106*(425), 557–586. https://doi.org/10.1093/afraf/adm059.

Chauveau, J.-P. (2006). How does an institution evolve? Land, politics, intergenerational relations and the institution of the tutorat amongst autochthones and immigrants (Gban region, Cote d'Ivoire). In R. Kuba & C. Lentz (Eds.), *Land and the Politics of Belonging in West Africa* (Vol. 9, pp. 213–240). Brill Academic Publishers. https://doi.org/10.1163/9789047417033_012.

Giovarelli, R., Hannay, L., Scalise, E., Richardson, A., Seitz, v., & Gaynor, R. (2015). *Gender and Land: Good Practices and Lessons Learned from Four Millennium Challenge Corporation Compact Funded Land Projects. Synthesis Report and Case studies: Benin, Lesoto, Mali, and Namibia*. Landesa and Center for Women's Land Rights. https://resourceequity.org/record/2734-gender-and-land-good-practices-and-lessons-learned-from-four-millennium-challenge-corporation-compact-funded-land-projects/.

Goldstein, M., Houngbedji, K., Kondylis, F., O'Sullivan, M., & Selod, H. (2015). *Formalizing Rural Land Rights in West Africa: Early Evidence from a Randomized Impact Evaluation in Benin*. https://doi.org/10.1596/1813-9450-7435.

Goldstein, M., Houngbedji, K., Kondylis, F., O'Sullivan, M., & Selod, H. (2018). Formalization without certification? Experimental evidence on property rights and investment. *Journal of Development Economics, 132*, 57–74. https://doi.org/10.1016/j.jdeveco.2017.12.008.

Institut National de la Statistique et de l'Analyse Économique (INSAE), & ICF. (2019). *Enquête Démographique et de Santé au Bénin (EDSB-IV) 2017–2018*. INSAE et ICF.

Lavigne Delville, P. (2019). *Chapter 7: History and Political Economy of Land Administration Reform in Benin*. Benin Institutional Diagnostic; Economic Development & Institutions.

Lavigne Delville, P., Colin, J.-P., Ka, I., & Merlet, M. (2017). Étude régionale sur les marchés fonciers ruraux en Afrique de l'Ouest et les outils de leur régulation (Vol. 1). Union Economique et Monétaire Ouest Africaine and Initiative Prospective Agricole et Rurale.

Lavigne Delville, P., & Moalic, A.-C. (2019). Territorialities, spatial inequalities and the formalization of land rights in Central Benin. *Africa, 89*(2), 329–352. https://doi.org/10.1017/S0001972019000111.

Le Meur, P.-Y. (2006). State making and the politics of the frontier in Central Benin. *D evelopment and Change, 37*(4), 871–900. https://doi.org/10.1111/j.1467-7660.2006.00505.x.

Platteau, J.-P. (2019). The spatial, historical and socio-political context. In F. Bourguignon, R. Houssa, J.-P. Platteau, & P. Reding (Eds.), *The Benin Institutional Diagnostic*. University of Namur. https://edi.opml.co.uk/resource/benin-the-spatial-historical-and-socio-political-context/.

World Bank Gender and Innovation Lab (WBGIL). (2019). *Impact Evaluation of Access to Land Project in Benin*. Prepared for the Millennium Challenge Corporation. https://thedocs.worldbank.org/en/doc/537351555943343180-0010022019/original/MCCEvaluationReportIEofAccesstoLandProjectinBeninFINAL.pdf.

Wren-Lewis, L., Becerra-Valbuena, L., & Houngbedji, K. (2020). Formalizing land rights can reduce forest loss: experimental evidence from Benin. *Science Advances, 6*(26), eabb6914. https://doi.org/10.1126/sciadv.abb6914.

Yemadje, R. H., Crane, T. A., Mongbo, R. L., Saïdou, A., Azontonde, H. A., Kossou, D. K., & Kuyper, T. W. (2014). Revisiting land reform: land rights, access, and soil fertility management on the Adja Plateau in Benin. *International Journal of Agricultural Sustainability, 12*(3), 355–369. https://doi.org/10.1080/14735903.2014.909645.

Yemadje, R. H., Crane, T. A., Vissoh, P. V., Mongbo, R. L., Richards, P., Kossou, D. K., & Kuyper, T. W. (2012). The political ecology of land management in the oil palm based cropping system on the Adja plateau in Benin. *NJAS: Wageningen Journal of Life Sciences, 60–63*(1), 91–99. https://doi.org/10.1016/j.njas.2012.06.007.

5 Ethiopia's Land Certification programs

Political economy context

The impetus for land certification in Ethiopia was the pervasive perception of tenure insecurity in rural areas, first under the imperial regime, followed by several decades of forced displacement linked to land redistribution and resettlement schemes initiated under the socialist (Derg) regime (1974–1987) and continued under subsequent political regimes (Holden et al. 2011). In 1975, the Government of Ethiopia nationalized all rural land and farmers were granted use rights, a policy that was enshrined in Ethiopia's 1995 Constitution and 1995 Land Proclamation (Holden et al. 2011). The Derg government periodically redistributed parcels to newly formed and landless households and prohibited sales, inheritance, and leasing (Benin & Pender 2001). However, restrictions on inheritance and leasing proved widely unpopular and these rights were reinstated in the post-Derg Land Proclamation of 1995 and in subsequent regional state proclamations. Amhara sought to implement a land redistribution in 1997 and 1998 that was also unpopular and was quickly abandoned. While discarding the unpopular and disruptive features of the Derg land reforms, the post-1995 national and regional reforms represented a strongly egalitarian approach to rural household land rights, based on the rights of social inclusion typical of customary tenure that prevailed in the pre-Derg era (Ayano 2018). All households were entitled to rights to land and pasture free of charge (analogous to the customary tenure principle of access to land as social right), duration of use rights have no limit, people secure land by donation or inheritance from family members (consistent with descent group-based customary principles) or from the state.

Program overview

Tigray started a land certification process in 1998–1999 using low-cost implementation methods that were later extended to other regions with donor support (Deininger et al. 2011; Holden et al. 2011). The program, which became known

DOI: 10.4324/9781003365679-5

as first-level land certification (FLLC), was expanded to Amhara in 2002 and to Oromia and the Southern Nations Nationalities and Peoples (SNNP) regions in 2004 (Cloudburst 2016). The FFLC programs, which were highly participatory and used a low-tech approach to parcel surveys, proved effective at providing tenure security for many certificate recipients (Holden et al. 2011).

The regional states generally adhered to a common set of procedures in implementing FLLC, including: building public awareness raising and establishment of an elected and independent village land use and administration committee (LAC); completion of a land registration application by landholders listing their land parcels, names of right holders, general location, land use, names of neighbors and, parcel boundary demarcation using measuring tapes, ropes, and field markings, in conjunction with the memories of the neighbors in bordering farms (Bezu and Holden 2014); public adjudication of claims by the LAC; entry of land rights into registry books kept at kebele and woreda levels, and issuance of a land certificate to heads of households, instead of by parcel (Ahmed 2017; Deininger et al. 2011; Melesse and Bulte 2015).

With USAID support through the Ethiopia Land Administration Program (ELAP) (2008–2013), the Ethiopian government implemented a program of second-level land certification (SLLC) in collaboration with woreda-level (district) land administration agencies in Tigray, Amhara, Oromia, and SNNP Regions. In the SLLC process, parcel boundaries are recorded with Global Positioning System (GPS) units and the data is entered into computerized land registries, making it easier for local land offices to compile land rights data and, in principle if not necessarily in practice, track land transactions (Cloudburst 2016). Farmers receive plot-level certificates with maps (Bezu and Holden 2014).

The multi-stranded program theory depicted in Tables 5.1 and 5.2 shaped the design of Ethiopia's first-level land and second level land certification programs. The top row of both tables lists contextual characteristics identified in the literature as the reasons for implementing a land registration process. Underneath the contextual factors, we list the different pathways that are theorized in the FLLC and SLLC program theory. For each theorized pathway, the subsequent rows describe its theorized mechanism, which consists of a resource and a response, followed by intermediate outcomes and the final outcome.

Realized outcomes

As indicated in Annex 5, the expectation that FLLC would improve tenure security was broadly supported (Ahmed 2017; Cloudburst 2016; de Brauw and Mueller 2012; Deininger et al. 2011; Holden et al. 2011; Kumar and Quisumbing 2015; Melesse and Bulte 2015; Yami and Snyder 2015).[2] An early evaluation indicated that tenure security gains of SLLC over FLLC were limited (Cloudburst 2016).[3] However, it is not possible to differentiate the

Table 5.1 CMO for increased agricultural productivity for First Level Land and Second Level Land Certification programs

1. Key tenure-related contextual factors identified by program designers as influencing agricultural productivity

- A long history of forced land redistribution and resettlement of rural smallholders created widespread fears that the government would expropriate farmland, discouraging investments in land that would improve productivity; documentation of use rights would alleviate those fears (Deininger et al. 2011; Holden et al. 2011).
- Many smallholders lacked state-issued documentation of land rights (Holden et al. 2011).
- Presence of land conflicts, which project designers attributed to the lack of state-issued documentation of land rights and absence of locally accessible land registries (Holden et al. 2011).
- Lack of rights documentation assumed to deter many rural households from engaging in land practices that increase agricultural productivity (Deininger et al. 2011; Melesse and Bulte 2015).
- In rural areas, women tend to have less security of tenure because they often have derived access and use rights to land and other resources (Ahmed 2017; Holden et al. 2011, Kumar and Quisumbing 2015).

2a. Theorized pathway following FLLC implementation for achieving investment in land management practices	**2b. Theorized pathway following FLLC implementation for achieving smoothing of land transactions**	**2c. Theorized pathway following FLLC implementation for access to credit**[1]
3a. Mechanism *Resource*: clear and enforceable land rights for landholders and users *Response*: improved perceived tenure security provides confidence for holders that their farm yields won't be disputed	**3b. Mechanism** *Resource*: clear and enforceable land rights for landholders *Response*: potential buyer or renter has assurance that the landholder has the right to transfer ownership or use rights, facilitating the sale and transfer of land to most productive farmers	**3c. Mechanism** *Resource*: clear and enforceable land rights for landholders *Response*: potential lender has assurance that the landholder has the right to transfer ownership rights; this enhances the collateral value of land, leading to easier access to credit for landholders
4a. Intermediate outcome 1 Greater investment in land management practices with long-term benefits (i.e., tree-planting, planting of perennial crops, construction of anti-erosion terraces, long-term fallows)	**4b-1. Intermediate outcome 1** Landholders who lack the capacity to farm the land can easily rent it out to someone who is able to farm it. FHH were expected to benefit from this mechanism the most given that social norms, as well as resource and time constraints limit their ability to use oxen to plough **Intermediate outcome 2** Renter with more assets can put the land into more efficient production	**4c-1. Intermediate outcome 1** Improved land market functioning (rental and sales) as third parties such as mortgage lenders easily access reliable land ownership information **Intermediate outcome 2** Landholders use borrowed funds to make investments in their land that improve soil fertility (i.e., application of improved fertilizers) or reduce erosion (i.e., tree-planting; construction of anti-erosion structures)

5. Final outcome: Agricultural productivity increases

Table 5.2 CMOs for reduced conflict and social inclusion for First Level Land Certification programs

1. Key tenure-related contextual factors identified by program designers as influencing land conflict and social inclusion	
• A long history of forced land redistribution and resettlement of rural smallholders created widespread fears that the government would expropriate farmland, discouraging investments in land that would improve productivity; documentation of use rights would alleviate those fears (Deininger et al. 2011; Holden et al. 2011)	
• Many smallholders lacked state-issued documentation of land rights (Holden et al. 2011)	
• Presence of land conflicts, which project designers attributed to the lack of state-issued documentation of land rights and absence of locally accessible land registries (Holden et al. 2011)	
• Lack of rights documentation assumed to deter many rural households from engaging in land practices that increase agricultural productivity (Deininger et al. 2011; Melesse and Bulte 2015)	
• In rural areas, women tend to have less security of tenure because they often have derived access and use rights to land and other resources (Ahmed 2017; Holden et al. 2011, Kumar and Quisumbing 2015)	
2a. Theorized pathway through FLLC implementation to reduced conflict	**2b. Theorized pathway through FLLC implementation to social inclusion**
3a. Farmer-to-farmer conflict	**3b. Access to land for women**
4a. Mechanism	**4b. Mechanism**
Resource: clear and enforceable land rights for landholders and users	*Resource:* clear and enforceable land rights for secondary rights holders (women and children) created through including names of female spouses on land certificates and targeting women in female headed households for certification
Response: potential rival land claimants believe their claim will not be upheld should they try to expropriate the land or encroach on boundaries; hence they have an incentive to respect borders	*Response*: women feel more secure, providing confidence that if they make investments in land that they or their heirs will benefit
5a. Intermediate outcome:	**5b. Intermediate outcomes:**
Fewer conflicts over land among farmers; less time devoted to guarding property and engaging in conflict, more time in productive activity including agriculture	Reduced eviction threats for married women upon death or divorce Equitable access to land for daughters and sons upon succession and inheritance Women participate in land use decisions
6a. Final outcome:	**6a. Final outcome**
Agricultural productivity increases	Women benefit from profits derived from land use and transferability of land; increased agricultural productivity

effects of SLLC over FLLC in a realist analysis. Many evaluations do not explain if they evaluated FLLC or SLLC, and over time, some of the effects due to FLLC alone not be identifiable since many households that originally received FLLC updated their certifications to SLLC.

Investment and productivity outcomes

Overall, certification has had a positive impact on smallholders' use of land improvement practices, such as soil and water conservation structures (Ahmed 2017; Alvarado et al. 2022; Deininger et al. 2011; Melesse and Bulte 2015), tree-planting (Melesse and Bulte 2015), and use of organic fertilizer (Melesse and Bulte 2015). Likewise, studies of the impacts of FLLC on agricultural productivity suggest positive outcomes (Melesse and Bulte 2015; Yami and Snyder 2015). However, more recent studies of SLLC do not find statistically significant productivity improvements (Alvarado et al. 2022).

Rental and credit activity

A particularly striking positive outcome of the certification process has been the increase in land rented or sharecropped out, particularly by female heads of household who lack access to male labor and other farming assets (Alvarado et al. 2022; Deininger et al. 2011; Holden et al. 2011; Yami and Snyder 2015). The FLLC program appears to have reduced transaction costs in the land rental market by making poor female (potential) landlord households more willing to rent out their land. It has, therefore, become easier for (potential) tenants to access land to rent in (Holden et al. 2011: 45). Certification, either FLLC or SLLC, also seems to have positively affected women's access to land (Cloudburst 2016). However, this mostly benefits women who are married or in unions, and not necessarily women who are divorced, widowed, or single (Alvarado et al. 2022). Holden et al. (2011) and Deininger et al. (2011) suggest that increased rentals enhance overall agricultural productivity since social norms and women's limited access to agricultural inputs likely would have left the land unused or underused.

FLLC appears to have had limited or modest impact on the use of credit (Yami and Snyder 2015). An evaluation of SLLC found a 10% increase in the use of informal credit (Cloudburst 2016). Land in Ethiopia cannot be used as collateral, so the mechanisms contributing to an increase in informal credit use are uncertain. However, Ethiopia has a robust rural informal credit market (Balana et al. 2022).

Contextual factors affecting outcomes

Annex 6 lists the contextual factors identified in the studies as contributing to PDR program outcomes. The widespread interest among rural Ethiopian landholders in land certificates and the generally positive outcomes of PDR are widely attributed to the country's recent prior history of disruptive land redistribution under the Derg regime from the mid-1970s to 1991 and fears of expropriation by the state due to abrogation of inheritance norms and violation of rules regulating fallowing (Deininger et al. 2011; Holden et al. 2011; Melesse and Bulte 2015; Yami and Snyder 2015). While Ethiopia's 1995 Constitution and 1997 Land Proclamation vest rural land ownership in the state and land cannot be sold, they explicitly protect inheritance rights, giving landholding families a long-term tenure horizon (Holden et al. 2011). Popular concerns about the kinds of disruptions in land rights associated with the Derg era have been put to rest. Noting that the tenure system in practice embodies key features of customary tenure, including prohibitions on land sales and membership in the local social group, it will be interesting to see if landholders feel it necessary to register future land transfers, including inheritance transfers, with local land authorities.

Land market impacts are largely restricted to changes in rental behavior. As noted in the description of outcomes, PDR-induced changes in rental and sharecropping behavior have benefited female headed households lacking access to male labor, with key contributing contextual factors in Tigray and Amhara being a high demand for land coupled with cultural norms that restrict women's use of oxen (Ahmed 2017; Deininger et al. 2011; Holden et al. 2011).

State and customary law: women's equity and access

Several of the studies suggest that the presence of legal frameworks supportive of gender equality for land rights at both federal and regional levels helps explain the relatively favorable social inclusion outcomes for Ethiopia's land certification program (Kumar and Quisumbing 2015; Lavers 2017; Melesse and Bulte 2015). Ethiopia's 1995 Constitution states that men and women have equal rights with respect to land access, management, transfers, and inheritance. Reforms to the federal Family Code in 2000 accorded equal rights to spouses upon marriage and divorce (Kumar and Quisumbing 2015); similar reforms were implemented in Tigray, Oromia, Amhara, and SNNP over the next five years. Women's rights to land were further strengthened with the issuance of the federal 2005 Land Proclamation, which required joint land registration (Kumar and Quisumbing 2015; Melesse and Bulte 2015). Kumar and Quisumbing (2015) argue that it is the combination of the 2000 Family Code and the 2005 Land Proclamation that has enabled women to benefit from certification, because together they provided a mutually reinforcing legal framework that promoted equality for women in land access and tenure security.

Any analysis that credits law reform alone for greater gender equality would overlook the relevance of cultural norms as important and enduring contextual factors. Law reforms supportive of gender equality are refracted through cultural norms according to which most women continue to access land through male relatives (Kumar and Quisumbing 2015; Lavers 2017). Regional differences in how the state legal framework intersects with local social norms may have led to positive outcomes for women in some areas and no impact on outcomes in others (Kumar and Quisumbing 2015; Lavers 2017). In Tigray, Lavers (2017) found that local tenure norms recognizing women's inheritance rights supported married women's informal claims to land even though they were not listed on certificates for land held in common with their husbands. In contrast, less supportive local tenure norms in Oromia made it difficult for women listed on joint titles to actualize their rights at their husband's death or upon divorce (Lavers 2017).

Enabling features of Ethiopia's certification programs

Several features of Ethiopia's large-scale land certification programs enhanced perceptions of tenure security. The initial FLLC program in Tigray, elements of which were taken up by other regions subsequently, emerged locally, was highly decentralized, relied on inexpensive and widely accessible mapping technology, and was implemented initially with limited donor involvement and by working through existing local governance institutions using a transparent process for demarcation and rights adjudication (Deininger et al. 2011; Holden et al. 2011). Arguably, locally led design and implementation of the Tigray program helped ensure that socially inclusive features of the customary tenure arrangements that predated the Socialist revolution would be taken up in ways less likely to occur when planning and design was in the hands of central government authorities or donors. Local design leadership was especially necessary given the political sensitivities associated with land reform in Ethiopia historically. Additionally, the program incorporated extensive awareness-raising campaigns to generate widespread knowledge of and interest in land certification (Deininger et al. 2011). Finally, and importantly, certification provided peasants with state-sponsored assurance that the disruptive land redistribution programs that characterized the Derg period would be less likely to return.

The center circle lists the key components of the FLLC programs in Ethiopia, while the outer circle lists factors that influenced the outcomes of interest in the three PDR cases included in this study. Block arrows link the two circles. The solid block arrows represent factors that played an important role in shaping Ethiopia's FLLC program outcomes. Block arrows that are not filled in indicate important factors that shaped outcomes in Benin and Rwanda, but which did not appear to be as critical in Ethiopia. The arrow labeled with a 1 indicates a male head of household who resides in a community where the FLLC program is being implemented; the arrow labeled with a 2 indicates a widowed female head of household who lives in the same community. The top portion of the box on the right describes the mechanism (i.e., resource and response) triggered by the FLLC program and the subsequent intermediate and

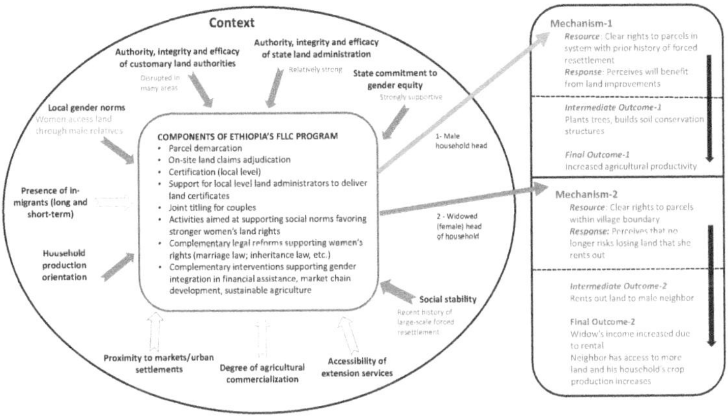

Figure 5.1 Illustrative revised CMO for Ethiopia's FLLC program explaining differential outcomes for gender.

final outcomes for male heads of households. The bottom portion describes the mechanism triggered by the FLLC program and subsequent intermediate and final outcomes for widowed female heads of households.

In this example, a male head of household has long-term heritable use rights to rural land under Ethiopia's system of state ownership. Through the FLLC program, the boundaries of the land he is using are clearly marked, neighboring farmers and elders have concurred that the land is his to use, and he now has a document that attests to his rights to use that land. The FLLC program therefore provides him with greater assurance that he will continue to have use rights indefinitely provided he continues to farm the land. Consequently, he decides to plant eucalyptus trees and construct terraces to reduce erosion. Although he is soon able to earn additional money from selling his eucalyptus trees to a local charcoal maker, it is too soon to tell whether his land has become more productive as a result of the erosion-control measures he's installed.

In the same village, a widowed female head of household also has long-term heritable use rights to rural land under Ethiopia's system of state ownership. However, under prevailing gender norms it would be very unusual for a woman to use oxen to plow her fields. Consequently, before she had the land certificate, her choices were to either sharecrop out her fields and risk losing them if the sharecropper brought a counterclaim to the land, or farm the land without oxen, which would greatly limit the size of the harvest. Having her land demarcated and her rights to use that land validated in a public process, together with written documentation of those rights, makes it clear to her and others in the community that she has the rights to that land. Having this clarity provides her greater assurance that a sharecropper is unlikely to be able to make a successful counterclaim to the land. The widowed female head of household therefore decides to sharecrop out her land rather than farming it herself. Consequently, her economic situation improves as she benefits from a share of the crop, as well as from non-farm income she earns now that she doesn't have to spend her time farming.

The illustrative CMO depicted in Figure 5.1 can be adapted to other contexts and other types of landholders within Ethiopia. The key point of the revised CMO is to illustrate the variety of factors that are likely to influence the resources that are provided by a tenure intervention, depending on a complex mix of contextual factors and individual or household characteristics. The resources (or lack of resources) provided by the tenure intervention in turn will affect the response and ultimately, the final outcomes of the intervention.

Notes

1 Although mortgages of rural land are prohibited in Ethiopia, certificate holders are securing informal credit, with the certificate seen by the lender as evidence of bankability, even though the land is not formally a loan security.

2 However, some scholars, including Ayano (2018) and Rahmato (2014), question how secure tenure is if one's rights are limited to use rights.

3 A preliminary report on outcomes of SLLC under the Land Investments for Transformation Program (LIFT), a land registration program supported by DFID and implemented by Development Alternatives Incorporated (DAI) has registered 14 million parcels, of which 70% belong to women. The program has invested heavily in public awareness and communications strategies designed to facilitate the registration of landholdings in the names of women and members of other vulnerable groups. These strategies, together with accompanying measures, such as facilitating smallholders' access to formal rental contracts and credit based on possession of a land certificate and strengthening local land administration capacity, are identified as important factors for improved outcomes for women landholders (DAI 2020).

References

Ahmed, H. S. (2017). *Gender and Rural Land Reform in Ethiopia: Reform Process, Tenure Security, and Investment* [Doctoral, University of Sussex]. http://sro.sussex.ac.uk/id/eprint/72414/.

Alvarado, G., Kieran, C., Jacobs, K., Beduhn, J., Heidenrich, T., & Linkow, B. (2022). *Ethiopia Strengthening Land Tenure and Administration Program Follow-on Report: An Impact Evaluation of Second-level Land Certification.* United States Agency for International Development (USAID). https://cdn.landesa.org/wp-content/uploads/ELTAP-ELAP-2021-Evaluation-Final_9.23.22.pdf.

Ayano, M. F. (2018). Rural land registration in Ethiopia: myths and realities. *Law & Society Review*, *52*(4), 1060–1097. https://doi.org/10.1111/lasr.12369.

Balana, B. B., Mekonnen, D., Haile, B., Hagos, F., Yimam, S., & Ringler, C. (2022). Demand and supply constraints of credit in smallholder farming: evidence from Ethiopia and Tanzania. *World Development*, *159*, 106033. https://doi.org/10.1016/j.worlddev.2022.106033.

Benin, S., & Pender, J. (2001). Impacts of land redistribution on land management and productivity in the Ethiopian highlands. *Land Degradation & Development*, *12*(6), 555–568. https://doi.org/10.1002/ldr.473.

Bezu, S., & Holden, S. T. (2014). Demand for second-stage land certification in Ethiopia: Evidence from household panel data. Land Use Policy, 41, 193–205. http://dx.doi.org/10.1016/j.landusepol.2014.05.013.

Cloudburst Group. (2016). *USAID Land Tenure ELTAP-ELAP Impact Evaluations Endline Report*. https://www.land-links.org/wp-content/uploads/2016/09/USAID_Land_Tenure_ELTAP-ELAP_Impact_Evaluations_Endline_Report.pdf.

de Brauw, A., & Mueller, V. (2012). Do limitations in land rights transferability influence mobility rates in Ethiopia? *Journal of African Economies*, *21*(4), 548–579. https://doi.org/10.1093/jae/ejs007.

Deininger, K., Ali, D. A., & Alemu, T. (2011). Impacts of land certification on tenure security, investment, and land market participation: evidence from Ethiopia. *Land Economics*, *87*(2), 312–334. https://doi.org/10.3368/le.87.2.312.

Development Alternatives Incorporated (DAI). (2020). *Fast Facts: Ethiopia: Land Investment for Transformation (LIFT) Programme (2013–2020)*. https://www.dai.com/uploads/Fast%20Facts%20Ethiopia%20LIFT%20Final%20040618v2.pdf.

Holden, S. T., Deininger, K., & Ghebru, H. (2011). Tenure insecurity, gender, low-cost land certification and land rental market participation in Ethiopia. *The Journal of Development Studies*, *47*(1), 31–47. https://doi.org/10.1080/00220381003706460.

Kumar, N., & Quisumbing, A. R. (2015). Policy reform toward gender equality in Ethiopia: little by little the egg begins to walk. *World Development*, *67*(C), 406–423.

Lavers, T. (2017). Land registration and gender equality in Ethiopia: how state–society relations influence the enforcement of institutional change. *Journal of Agrarian Change*, *17*(1), 188–207. https://doi.org/10.1111/joac.12138.

Melesse, M. B., & Bulte, E. (2015). Does land registration and certification boost farm productivity? Evidence from Ethiopia. *Agricultural Economics*, *46*(6), 757–768. https://doi.org/10.1111/agec.12191.

Rahmato, D. (2014). The perils of development from above: land deals in Ethiopia. *African Identities*, *12*(1), 26–44. https://doi.org/10.1080/14725843.2014.886431.

Yami, M., & Snyder, K. (2015). After all, land belongs to the state: examining the benefits of land registration for smallholders in Ethiopia. *Land Degradation & Development*, *27*. https://doi.org/10.1002/ldr.2371.

6 Rwanda's Land Tenure Regularization program

Program overview

Historical inequalities in land access and ownership in Rwanda contributed significantly to the 1994 Genocide (Andre and Plateau 1998). The return of several hundred thousand refugees after the Genocide and the ascension of the Kagame government further exacerbated land conflict (Andre and Plateau 1998). The post-Genocide government implemented land-related legal reforms as well as a country-wide Land Tenure Regularization (LTR) program to reduce land-related conflicts, address gender inequalities in access to land, and foster economic development (Ali et al. 2014). The LTR program documented existing use rights to the country's ten million plus land parcels and was viewed as an intermediate step toward eventual legal recognition through the issuance of a title to the landholders (Ali et al. 2014). LTR was designed to establish the right to private property as inviolable, encourage efficient use of land resources and eliminate discrimination in access to land (Santos et al. 2014). Through this low-cost, locally based implementation at the sector and cell levels[1] and participatory approach,[2] informal customary rights for more than 8.4 million land parcels were converted into formally registered rights in three years (Bayisenge 2018).[3] Rwanda's 2016 land administration procedures manual indicates that customary tenure continued to exist alongside formally registered land.

Prior to the 2005 Land Law, which authorized the LTR program, Rwanda's 2004 National Land Policy set the stage for two programs, the Crop Intensification Program (CIP) and the Land Use Consolidation Program (LUC) intended to catalyze intensification of agricultural land use. The LUC initially encouraged farmers to consolidate their landholdings and provided subsidized CIP inputs and other forms of support only to holdings one hectare or larger. By clarifying and registering land ownership, it was hoped that farmers would use their title as collateral for loans to finance agricultural inputs, improvements, innovations, and expansion of their enterprises (Muyombano et al. 2018).

As indicated in Tables 6.1 and 6.2, Rwanda's LTR program consisted of multiple non-exclusive strands. The top row of both tables lists contextual characteristics identified in the literature as the reasons for implementing a

DOI: 10.4324/9781003365679-6

land registration process. Underneath the contextual factors, we list the different pathways that are theorized in the LTR program theory. For each theorized pathway, the subsequent rows describe its theorized mechanism, which consists of a resource and a response, intermediate outcomes, and the final outcome.

Table 6.1 CMOs for increased agricultural productivity for the Land Tenure Regularization program

1. Key tenure-related contextual factors identified by program designers as influencing agricultural productivity • Severe conflict over land rights and landlessness driven by widespread informal sales were contributing factors to the 1994 Genocide, creating a need for clarifying customary use rights (Andre and Platteau 1998) • Many smallholders lacked state-issued documentation of land rights (Ali et al. 2014) • Presence of land conflicts which were assumed to be exacerbated by the lack of state-issued documentation of land rights and absence of locally accessible land registries (Ali et al. 2014) • Lack of rights documentation was assumed to deter many rural households from engaging in land practices that increase agricultural productivity (Muyombano et al. 2018) • In rural areas, women tended to have less security of tenure because they often have derived access and use rights to land and other resources (Ali et al. 2014)		
2a. Theorized pathway following LTR implementation for achieving investment in land management practices	**2b. Theorized pathway following LTR implementation for facilitating of land transactions**	**2c. Theorized pathway following LTR implementation for improved access to credit**
3a. Mechanism *Resource*: clear and enforceable land rights for landholders and users *Response*: improved perceived tenure security provides confidence for holders that their farm yields won't be disputed.	**3b. Mechanism** *Resource*: clear and enforceable land rights for landholders *Response*: potential buyer or renter has assurance that the landholder has the right to transfer ownership or use rights, facilitating the sale and transfer of land to most productive farmers	**3c. Mechanism** *Resource*: clear and enforceable land rights for landholders *Response*: potential lender has assurance that the landholder has the right to transfer ownership rights; this enhances the collateral value of land, leading to easier access to credit for landholders and potential buyers
4a. Intermediate outcome 1 Greater investment in land management practices with long-term benefits (i.e., tree-planting, planting of perennial crops, construction of anti-erosion terraces, long-term fallows)	**4b-1. Intermediate outcome 1** Landholders who lack the capacity to farm the land can easily rent it out or sell it to someone who is able to farm it. FHH were expected to benefit from this mechanism the most given that social norms, as well as resource and time constraints, limit their ability to use oxen to plough **4b-2. Intermediate outcome 2** Purchaser or renter with more assets can put the land into more efficient production	**4c-1. Intermediate outcome 1** Improved land market functioning (rental and sales) as third parties such as mortgage lenders easily access reliable land ownership information **4c-2. Intermediate outcome 2** Landholders use borrowed funds to make investments in their land that improve soil fertility (i.e., application of improved fertilizers) or reduce erosion (i.e., tree-planting; construction of anti-erosion structures)
5.0 Final outcome: agricultural productivity increases		

Table 6.2 CMOs for reduced conflict and social inclusion for the LTR program

1. Key tenure-related contextual factors identified by program designers as influencing land conflict and social inclusion

- Severe conflict over land rights and landlessness driven by widespread informal sales were contributing factors to the 1994 Genocide, creating a need for clarifying customary use rights (Andre and Platteau 1998)
- Many smallholders lacked state-issued documentation of land rights (Ali et al. 2014)
- Project designers assumed that the lack of state-issued documentation of land rights and absence of locally accessible land registries exacerbated land conflicts in much of the country (Ali et al. 2014)
- Project designers assumed that insecure tenure associated with lack of rights documentation had deterred many rural households from engaging in land practices that increase agricultural productivity (Muyombano et al. 2018)
- In rural areas, women tend to have less security of tenure because they often have derived access and use rights to land and other resources (Ali et al. 2014)

2a. Theorized pathway through LTR implementation to reduced conflict	**2b. Theorized pathway through LTR implementation to social inclusion**
3a. Farmer-to-farmer conflict Mechanism	**3b. Access to land for women Mechanism**
Resource: clear and enforceable land rights for landholders and users *Response*: potential rival land claimants believe their claim will not be upheld should they try to expropriate the land or encroach on boundaries; hence they have an incentive to respect borders	*Resource:* clear and enforceable land rights for secondary rights holders (women and children) created through including names of female spouses on land certificates and targeting women in female headed households for certification *Response*: women feel more secure, providing confidence that if they make investments in land that they or their heirs will benefit
Intermediate outcome: Fewer conflicts over land among farmers; less time devoted to guarding property and engaging in conflict, more time in productive activity including agriculture	**Intermediate outcomes:** Reduced eviction threats for married women upon death or divorce Equitable access to land for daughters and sons upon succession and inheritance Women participate in land use decisions
Final outcome: *Agricultural productivity increases*	**Final outcome:** *Women benefit from profits derived from land use and transferability of land; agricultural productivity increases*

Realized outcomes

Outcomes for Rwanda's LTR program are listed in Annex 5. Research suggests that the LTR program enhanced tenure security (Abbott et al. 2018), with differential gains and losses among categories of rural residents. Details of the outcomes are included in Annex 5.

Women's outcomes

The outcomes for women were mixed, and studies found contradictory results. Ali et al. (2014: 274) reported for the pilot phase that "legally married women were significantly more likely to have their informal rights documented and secured," but that the rights of female spouses who were not in legally recognized marriages were eroded. However, in a study post-implementation, Santos et al. (2014) found that wives in polygamous marriages not recognized as legal marriages under the 1999 Matrimonial Regimes, Liberties, and Succession Law (MRLSL), were registering their plots in their own names and registering their children as heirs. Although husbands continued to have control over the plots, women in polygamous marriages were more likely to perceive gains in tenure security than women legally married in terms of the MRLSL (Santos et al. 2014: 39). Boundary disputes decreased post-implementation, but Muyombano et al. (2018) and Santos et al. (2014) found that disputes over inheritance had increased. Fears on the part of men or women that the state might expropriate land do not appear to have been reduced (Abbott and Mugisha 2015; Ali et al. 2014).

Investment outcomes

Outcomes for investments in land improvements were mixed. Ali et al. (2014) and Bayisenge (2018) documented an increase in soil conservation investments, particularly among women. However, Abbott and Mugisha (2015) found no increase in the amount of land with soil erosion protection structures, although they did find increases in the use of soil fertility enhancements, which they attributed to fertilizer subsidies provided for plots larger than one-hectare, a provision of the LUC. Abbott and Mugisha (2015) found no impact of the LTR on the rate of increase in agricultural productivity.

Loans and land transactions

Muyombano et al. (2018) and Abbot and Mugisha (2015) both reported that a small percentage of LTR program beneficiaries used their titles to secure loans. Citing a 2012 study, Abbot and Mugisha (2015) report that loans were principally used for housing construction and only about 6% of loans were used for purchasing farm inputs.

Reported outcomes for land transactions were mixed: Ali et al. (2014) reported a decrease in land sales, but Abbot and Mugisha (2015) and Muyombano et al. (2018) found no impact on land sales. Muyombano et al. (2018) reported that rentals had increased.[4]

Ali et al. (2019) found active informal market transactions in land in rural areas, but low rates of registration of transfers—five years after the start of the LTR program 87% of land transactions went unregistered—raising concerns about erosion of tenure security of parcels registered in the initial universal registration of holdings. Based on survey data, Ali et al. (2019) conclude that landholders would be more likely to register permanent land transfers if the costs of doing so were lower. The authors appear not to consider the possibility that low levels of registration of transactions might be because perceived levels of tenure security were high, since transactions for the most part took place between relatives and other fellow members of the rural social system, backed up by assurances provided by the local customary tenure norms (Brown and Hughes 2017). Ali et al. (2014) found in a survey of participants in a LTR pilot project that households that had bought or inherited their land felt less insecure than those who had been allocated land by the government.

Contextual factors affecting outcomes

This section summarizes the key contextual key factors affecting the outcomes of Rwanda's LTR program. A more detailed picture of these factors is provided in Annex 6.

Women and enabling legal framework

Several evaluations (Abbott and Mugisha 2015; Abbott et al. 2018; Bayisenge 2018; Santos et al. 2014) attribute the generally positive outcomes for women's land rights to an enabling legal framework—a combination of land, marriage, and inheritance laws—which mainstreams gender in government policies and imposes gender quotas in governance and mandates gender-responsive budgeting (Abbott et al. 2018). The MRLSL of 1999 gave equal rights to daughters and sons to inherit their parents' property as well as equal rights to married women to inherit land from both their birth family and their husbands' family (Bayisenge 2018). However, inheritance rights are restricted to the family members of a civil marriage and to children born out of wedlock formally recognized by a parent during their lifetime (Abbott and Mugisha 2015). The marriage law provides no protection to women or men living in consensual unions or their children (unless they have been officially recognized by their parents during their lifetime) (Abbott and Mugisha 2015; Abbott et al. 2018). However, there is nothing in the land law that prevents partners in consensual unions from registering land individually. Women in polygamous marriages registered plots they farmed in their names only and registered their children as heirs and registered their husbands as parties of interest (Abbott et al. 2018). As a result, Santos et al. (2014) concluded that women who had land titled in their name only were more likely to feel secure than those who had joint titles with their spouse.

The 2005 Organic Land Law (OLL) provided for joint titles for couples whose marriages are registered under the MRLSL of 1999 (Bayisenge 2018). A principal aim of joint registration was to reduce the threat of eviction of widows in the event of separation or death of a spouse. While Bayisenge (2018) indicates that joint ownership of property is conditional on the marriage being legally registered under the MRLSL of 1999, Daley et al. (2010: 135) argue that the English version of the OLL describing joint marital ownership in terms of "husband and wife" is a mistranslation of the original Kinyarwanda version of the law, which provides that "the man and the woman have equal rights over the land, and not just the husband and wife." The mistranslation led to widespread misinterpretation on the part of program implementers that women in common law marriages (and therefore whose marriages were not recognized by the state) should not be included as joint landholders. A seemingly gender-neutral restriction of the OLL prohibiting the division of landholdings smaller than one hectare further affected women because when families are limited to choosing one heir, they are more likely to choose a man (Bayisenge 2018; Santos et al. 2014). However, a change in this provision is in process.

Customary norms and disadvantaged women

The continuance of customary social norms has contributed to LTR's variable gender outcomes. Abbott and Mugisha (2015) note that few men or women have a detailed understanding of the land and inheritance laws. Village leaders and the *Abunzi* (voluntary mediators) at cell and sector level who are responsible for arbitrating land disputes have insufficient training and do not have access to resources needed to support them in their work. In such circumstances, customary law often prevails, hindering women's ability to exercise their statutory land rights (Abbott and Mugisha 2015; Bayisenge 2018; Muyombano et al. 2018). In some instances, women who are formally joint landowners may acquiesce to their husbands in decision-making to avoid domestic conflicts (Bayisenge 2018).

Bayisenge (2018) found that less well-off, less-educated, older women who did not participate actively in community activities, as well as women who were not in leadership positions and, in some instances, those in unregistered marriages, had less knowledge of their land rights and were less able to claim their rights. They also found that women in very poor households were less likely to be listed on land titles and their daughters were less likely to inherit land (Bayisenge 2018). Addressing such differences will likely require grassroots-level behavioral change campaigns, in addition to sensitization of officials, legal aid providers, and other service providers to the needs of the poorer households (Santos et al. 2014).

Policy: fragmentation, subsidies, appropriation, and registration barriers

Provisions in the LUC and CIP programs to discourage fragmentation have not contributed to higher levels of productivity. Rather, Ali et al. (2014), drawing from a survey of participants in an LTR pilot program, find that it is the renewed social stability associated with LTR that has enabled the success of programs aimed at increasing agricultural productivity and reducing land fragmentation. Gains observed in priority commercial crop production and a corresponding decline in the production of staple crops are attributed to the provision of subsidized inputs and other support to farmers on consolidated land (Abbott and Mugisha 2015).

Additionally, under the OLL and the Land Policy, all land ultimately belongs to the government, with citizens holding 99-year leases. The state's power to take back and reallocate land that it deems improperly used or needed for infrastructure projects continues to fuel fears of expropriation, especially near urban areas (Abbott et al. 2018). As noted above, Ali et al. (2019) attributed the low level of formal registration of transactions to the excessive price of registration relative to the value of land, particularly for rural land. Brown and Hughes (2017) found that low levels of formal registration of transactions is linked to the illegality of subdivisions of less than one hectare. However, Brown and Hughes (2017) also reported that many of the unregistered transactions consist of inheritances or inter vivos gifts (umunani) between parents and children. These transactions are typically witnessed and validated by friends, family members, and neighbors, a practice that is consistent with Rwanda's customary tenure norms. Brown and Hughes (2017: 17) noted that village leaders provide additional security by witnessing land sales and extra-legal subdivisions.

Revised CMO for Rwanda: an illustrative example

Figure 6.1 depicts an alternative CMO for Rwanda PFR activity that seeks to capture some of its complexities.

The center circle lists the key components of Rwanda's LTR program, while the outer circle lists factors that influenced the outcomes of interest in all three of the PDR cases included in this study. Block arrows link the two circles. The solid block arrows represent factors that played an important role in shaping the LTR program's outcomes. Block arrows that are not filled in indicate important factors that shaped outcomes in Benin and Ethiopia, but which did not appear to be as critical in Rwanda. The arrow labeled with a 1 indicates a well-off woman in a state-recognized marriage who resides in a community where LTR is being implemented; the arrow labeled with a 2 indicates a poor woman in a state-recognized marriage who lives in the same community. The top portion of the box on the right describes the mechanism (i.e., a resource and response) triggered by the FLLC program and the subsequent intermediate and final outcomes for the wealthy women in a state-recognized marriage. The bottom portion describes the mechanism triggered by the FLLC program and subsequent intermediate and final outcomes for a poor woman in a state-recognized marriage.

In this example, the wealthy woman who is in a state-recognized marriage is from a very powerful local family and is a leader in her community. She is well aware of the LTR program and the requirement that land held by couples should be jointly titled. Consequently, when the land that she and her husband own together is being demarcated and registered, she makes sure that her name is

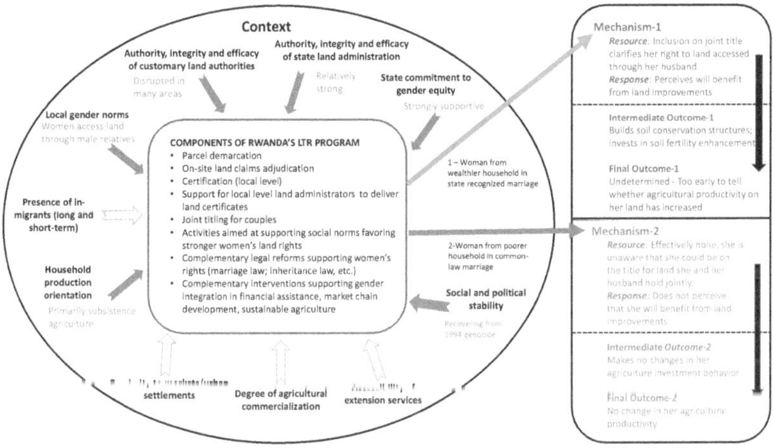

Figure 6.1 Illustrative revised CMO for Rwanda's LTR program depicting differential outcomes for women depending on their wealth status.

included on the title. The LTR program therefore provides her greater assurance that she will continue to have rights to the parcel even if her husband dies or she gets a divorce. Consequently, she hires her neighbor to build terraces to reduce erosion on the portion of the jointly owned land that she farms. It is too soon to tell whether her land has become more productive. In the same village, a poor woman who is in a common-law marriage is in a much different situation. She has no time to participate in local meetings and her social network does not include any community leaders or educated women who know about the LTR program. Consequently, she is unaware that the law requires jointly held land to be titled in both the husband and wife's name, and her name does not appear on the title that her common-law husband obtains through the LTR program. She knows that in her community, the custom is that the land she has access to through her husband will revert to his family upon his death or divorce. As a result, she is uncertain whether she would benefit from making any long-term investments in the land, and she can't afford to buy enhanced fertilizers to improve yields. Consequently, she makes no changes in her land investment behavior and the productivity of the land she farms remains the same.

The illustrative CMO depicted in Figure 6.1 can be adapted to other contexts and other types of landholders within Rwanda. As with the revised CMOs in Chapters 4 and 5, Rwanda's revised CMO serves primarily to illustrate the variety of factors that are likely to influence the resources provided by a tenure intervention, depending on a complex mix of contextual factors and individual or household characteristics. The resources (or lack of resources) provided by the tenure intervention in turn will affect the response and ultimately, the final outcomes of the intervention.

Notes

1 Cells are the lowest administrative units in Rwanda; sectors are the next level, followed by districts and finally provinces.
2 Parcel demarcation and rights validation involved the landholder, holders of neighboring parcels, as well as local authorities.
3 Rwanda's 2016 land administration procedures manual indicates that customary tenure continued to exist alongside formally registered land, with the latter being located primarily in urban areas.
4 Ali et al. (2014) included a counterfactual since they examined pilot results. The Abbot and Mugisha (2015) and Muyombano et al. (2018) studies were completed after LTR was completed and most land was registered, so no counterfactual was possible or scientifically necessary.

References

Abbott, P., & Mugisha, R. (2015). *Land Tenure Regularisation Programme: Progress Report for Selected Indicators*. United Kingdom Department for International Development (DfID).

Abbott, P., Mugisha, R., & Sapsford, R. (2018). Women, land and empowerment in Rwanda. *Journal of International Development*, *30*, 1006–1022. https://doi.org/10.1002/jid.3370.

Ali, D. A., Deininger, K., & Goldstein, M. (2014). Environmental and gender impacts of land tenure regularization in Africa: pilot evidence from Rwanda. *Journal of Development Economics*, *110*, 262–275. https://doi.org/10.1016/j.jdeveco.2013.12.009.

Ali, D. A., Deininger, K., Mahofa, G., & Nyakulama, R. (2019). Sustaining land registration benefits by addressing the challenges of reversion to informality in Rwanda. *Land Use Policy*, 104317. https://doi.org/10.1016/j.landusepol.2019.104317.

Andre, C., & Platteau, J.-P. (1998). Land relations under unbearable stress: Rwanda caught in the Malthusian trap. *Journal of Economic Behavior & Organization*, *34*(1), 1–47. https://doi.org/10.1016/S0167-2681(97)00045-0.

Bayisenge, J. (2018). From male to joint land ownership: women's experiences of the land tenure reform programme in Rwanda. *Journal of Agrarian Change*, *18*(3), 588–605. https://doi.org/10.1111/joac.12257.

Brown, M., & Hughes, A. K. (2017, March 20). *Is Land Tenure "Secure Enough" in Rural Rwanda?* 2017 World Bank Conference on Land and Poverty, Washington, DC.

Daley, E., Dore-Weeks, R., & Umuhoza, C. (2010). Ahead of the game: land tenure reform in Rwanda and the process of securing women's land rights. *Journal of Eastern African Studies*, *4*(1), 131–152. https://doi.org/10.1080/17531050903556691.

Muyombano, E., Espling, M., & Pilesjo, P. (2018). Effects of land titling and registration on tenure security and agricultural investments: case of Gataraga sector, Northern Rwanda. *African Journal on Land Policy and Geospatial Sciences*, *1*(3), Article 3. https://doi.org/10.48346/IMIST.PRSM/ajlp-gs.v1i3.14424.

Santos, F., Fletschner, D., & Daconto, G. (2014). Enhancing Inclusiveness of Rwanda's Land Tenure Regularization program: insights from early stages of its implementation. *World Development*, *62*, 30–41. https://doi.org/10.1016/j.worlddev.2014.04.011.

7 Zimbabwe's Fast Track Land Reform program

Program overview: rectifying a colonial legacy of land dispossession

Zimbabwe achieved its independence from Great Britain in 1980 after a long war between the white settler community, led by a breakaway government that illegally declared independence from Great Britain in 1965, and African national liberation movements. No country, including Great Britain, had recognized the white-led Rhodesian government's unilateral declaration of independence. Under colonial rule, wealth, land, and power had been concentrated in the hands of the white minority, constituting less than 5% of the population. The 1930 Land Apportionment Act divided rural land along racial lines, with 50,000 whites receiving 50% of the land and one million Africans confined to living in so-called Tribal Trust Lands, commonly referred to as communal areas, constituting about 30% of the national territory (Mlambo 2014). In light of the history of colonial land theft and political, economic, and social marginalization, restitution and/or redistribution of land rights to African citizens residing in rural areas were central political and policy concerns of the newly established majority-rule government that took power in 1980 (Aliber and Cousins 2013; Hull et al. 2019).

Great Britain brokered the Zimbabwe independence negotiations, which resulted in the Lancaster House Agreement that committed Great Britain to help finance a post-independence land redistribution program. Purchases of white-owned farmland for resettlement of African farmers would be negotiated on a willing seller—willing buyer basis. Up until 1999, funding made available for commercial farm purchase in terms of the Lancaster House Agreement helped transfer about 3 million hectares of the 15 million hectares held by white farmers to African beneficiaries (Scoones et al. 2011). Zimbabwe's Fast-Track Land Reform Program (FTLRP), launched in 2000, was designed to speed up the land redistribution process by initially identifying up to five million hectares of land for compulsory acquisition and demarcation (Moyo 2011). FTLRP proceeded on the basis of land acquisition by the

DOI: 10.4324/9781003365679-7

state without compensation to white farmers. White farmers held their land under title deed; all land confiscated and transferred to black farmers was held as state property, with farmers holding land under permits or leases. The European Union (led by Great Britain) and the US condemned state acquisition of land without compensation as violations of the human rights of white commercial farmers, and imposed targeted sanctions on the Zimbabwe government, restricting financial support for implementation of the FTLRP program.

The FTLRP program supported two intervention models. Model A1 aimed to decongest the overcrowded communal areas and assist subsistence farmers with limited land (Zikhali 2010). Beneficiaries of A1 land allocations were initially to have received offer letters from the government, essentially granting permission to resettle. These were to be followed by permits, as evidence of state validation of permanent land occupancy rights. A2 areas were designed as resettlement schemes for commercially oriented medium- to large-scale farmers (with holdings of 30 to 100 hectares), based on the principle of full-cost recovery from the beneficiaries. A2 farmers were to receive state-issued 99-year leases (Zikhali 2010). Unlike the previous two phases of Zimbabwe's post-independence land reform, where land for redistribution was to be purchased from white commercial farmers on a "willing seller-willing buyer" basis, the FTLRP confiscated white-owned commercial farmland without compensation (Mkodzongi and Lawrence 2019: 2). By 2008–2009, over 4,500 farms constituting 7.6 million hectares, 20% of the total land area of the country, had been distributed to over 145,000 farm households in A1 schemes and around 16,500 households in A2 schemes, equivalent to 11.7% of farm households in the country (Scoones et al. 2011).

The program theory underlying FTLRP in Zimbabwe stands on four pillars:

1 Restorative justice, principally by making more land available to black Zimbabweans from urban and rural areas, including from overcrowded communal areas;
2 Restructuring of the agrarian economy, with African majority participation across farm-sizes and agriculturally linked businesses;
3 Improving LRD beneficiary investment, leading to higher productivity and incomes; and
4 An orientation toward market production, particularly on larger-scale A2 farms whose new owners showed promise of being successful entrepreneurial farmers.

The theorized mechanisms and outcomes for Zimbabwe's FTLRP land redistribution program and its four pillars are described in Table 7.1

Table 7.1 CMOs (program theory) for Zimbabwe's FTLRP land redistribution program

2.0 Land redistribution program design: core assumptions about the intervention's context

- A history of forced displacement and land expropriation (Aliber and Cousins 2013)
- High levels of rural poverty due to limited access to land
- Smallholders have inadequate access to inputs needed for commercial production (Zikhali and Chilonda 2012)
- State authorities recognize two categories of reform beneficiary: A1 smallholder subsistence farmers and A2 commercially oriented farmers. The state undertakes to provide A1 farmers with "permits" as evidence of right to occupy land and A2 farmers with 99-year leases
- Authority and legitimacy of traditional land authorities had been challenged in the initial decade after independence, but these authorities gained legal recognition and political legitimacy before and during FTLRP implementation

2.0 Theorized pathways to increased equity and productivity following land redistribution

3a. Redistributive effect	**3b. Investment effect**	**3c. Market effect**[a]
4a. Mechanism	**4b. Mechanism**	**4c. Mechanism**
Resource: farming land made available to families from overcrowded communal areas along with subsidized fertilizers and access to credit *Response*: beneficiaries perceive that additional land and input supports will enable increases in farm-based incomes	*Resource*: clear and enforceable land rights for landholders documented in state-issued rights certificates *Response*: landholders perceive tenure is secure	*Resource*: clear and enforceable land rights for landholders coupled with provision of credit and subsidized fertilizer *Response*: landholders feel that they have the resources they need to shift from subsistence production orientation to commercial orientation
5a. Intermediate outcome Landless and land-poor have greater access to land with higher agricultural productivity potential. Landholders apply household labor to agriculture and make investments in their land that improve soil quality	**5b. Intermediate outcome** Landholders invest in soil improvements, fertilizer, and other inputs	**5c-1. Intermediate outcome 1** Labor shifted from non-farm to on-farm activities **5c-2. Intermediate outcome 2** Landholders make investment in their land that improve soil quality and productivity
6a. Final outcomes More equitable distribution of agricultural land that would permit sustainable and secure farm-based livelihoods. Agricultural production increases relative to that of communal land	**6b. Final outcomes** Agricultural production increases Farm-based incomes increase	**6c. Final outcomes** Agricultural productivity increases Farm-based income increases relative to non-farm income. Restorative justice achieved for large numbers of historically displaced people

a FTLRP resettlement land belongs to the state. A1 smallholders are meant to have use permits, A2 "commercial" farmers are meant to have 99-year leases. Most have neither yet. No sales are allowed.

Realized outcomes

Land invasions

A striking and unplanned feature of FTLRP implementation was land invasions by African farmers of white commercial farming areas earmarked for eventual redistribution to land reform beneficiaries. According to Scoones et al. (2011: 971), Most A1 sites were occupied by organized land invasions beginning in 2000. The settlers originated from surrounding communal areas and nearby towns and were led by war veterans and traditional authorities (Scoones et al. 2011: 971). Government policymakers and local officials were left flat-footed, unprepared to provide planning, logistical, and administrative support to the settlers. A key question of any land redistribution program, how tenure security would be delivered, had not been thought out or agreed upon within the government in advance of the land invasions (Matondi 2012: 16). According to Scoones et al. (2011: 972), issuance of "offer letters" and more formal demarcation of plots was left to be determined later. Offer letters, though, do not provide evidence of legal ownership, but rather indicate the state's provisional intention to formalize rights at a future date once stipulated conditions are met.

By 2011, the total area of land redistributed reached over seven million hectares (Scoones 2011), exceeding by good measure the government's initial goal of five million hectares. Though successful in redistributing land to thousands of beneficiaries, FTLRP implementation has been characterized by policy disputes over tenure policy and administrative inefficiencies in delivering tenure documentation to beneficiaries. Smallholder A1 farmers were expected to receive permits as evidence of long-term occupation rights, but less than 10% of promised A1 permits had been issued by the end of 2014 (Nyoni 2016: 12). A2 farmers were to have received 99-year leases, but issuance of leases is limited.

Tenure security

Consensus about how best to deliver land tenure security to land reform beneficiaries remains elusive. Many in the government argue for keeping FTLRP land under state ownership and issuing permits to A1 smallholders and 99-year leases to mid-to large-scale farmers. Matondi (2012: 99) points out that state ownership of land ensures a measure of state political control over the rural population, noting that the state had been careful not to leave the impression that it was prepared to give up its own rights in favor of strengthening individual rights. A survey cited by Matondi (2012) of landholders in Mazowe District showed that 55% of A1 farmers and 51% of A2 farmers preferred to have title deeds, and 42% of A2 farmers wanted 99-year leases. In Matondi's view, the majority of farmers were expressing a preference for what they perceived as a form of secure ownership that also enabled them to use land as collateral (Matondi 2012: 127).

Early in the reform process, traditional authorities sought restitution of historic rights to land in FTLRP areas which their ancestors administered before being displaced by the British. Several claims were successful. The 2013 Constitution granted significant powers to traditional authorities in local governance, including powers over land administration in communal areas, and the Traditional Authorities Act stipulates that all resettlement areas should be placed under the relevant chiefs or headman (Mkodzongi 2016: 100). Across rural Zimbabwe, traditional authorities and district councils share power and are required to consult on a variety of matters, including matters related to land administration and land use governance (Scoones et al. 2011). Offer letters are still the principal evidence of tenure over land held by A1 FTLRP beneficiaries (Nyoni 2016). Large numbers of holdings are unsurveyed, and village chiefs, not local government officials, adjudicate boundary and other disputes (Mkodzongi 2016).

While the government largely failed to deliver state-backed documentation of tenure apart from the initial letters of offer, there was a resurgence of the roles of traditional authorities in rural life, including in land matters, beginning in the late 1990s and gaining momentum in the early years of FTLRP implementation. Traditional authorities had been discredited by the Zimbabwe African National Union—Patriotic Front (ZANU-PF) government in the first decade after independence for presumed complicity with the system of indirect colonial rule. These views softened as traditional authorities continued to demonstrate legitimacy among many of their rural constituents, something that had eluded Village Development Committees (VIDCOs) and Rural Development Committees (RDCs), local-level civil bodies that were established by the central government (Mkodzongi 2016).

While VIDCOs were meant to provide for rural democratic decision-making at the local level, they lacked real authority, relying on the central government for their funding and key decisions, undermining their ability to discharge their duties (Mkodzongi 2016). In fact, the central government's attempts to establish hegemony in the countryside remained tenuous at best. Emerging from the competition between central government surrogates and customary authorities was a melange of institutions—VIDCOs and RDCs on the one hand, and customary authorities on the other—vying for the authority to allocate land. Mkodzongi (2016) observes that though land allocation powers had been officially transferred from chiefs to the VDCs (where chiefs sat as ex-officio members) chiefs remained popular among community members who continued to recognize their land allocation powers.

Investment and productivity considered in the context of a changing agrarian structure

Studies of the first ten years of FTLRP found evidence of investment and productivity declines across several types of crops when compared to levels on the white-owned commercial farms in the period before implementation of the program. Between 2001 and 2009, production of wheat, tobacco, coffee, and tea declined in areas acquired by FTLRP, as did beef exports (Scoones et al. 2011: 972). Zikhali and Chilonda (2012), citing 2011 World Bank data, found that national agricultural production declined by close to 58% between when the FTLRP was initiated in 2000 and 2009.

Muchetu (2019) argues that investment and productivity outcomes could have been better with greater state and private financial support for the newly resettled farmers. Muchetu (2019: 51) found that the share of public investment allocated to agriculture was low, averaging less than 5% of the national budget and with subsidies arriving late. FTLRP farm productivity was higher where farmers had access to the kinds of subsidies enjoyed by white commercial farmers. Zikhali and Chilonda (2012) evaluated the effects of fertilizer use on FTLRP holdings in Mazowe District, where FTLRP land allocations were accompanied by subsidized or preferential support for agricultural inputs. They found higher levels of productivity by FTLRP beneficiaries than for the control group of communal land farmers who did not receive preferential access to inputs. Surveys of FTLRP farmers carried out in 2006–2007 and 2013–2014 found that less than 10% considered tenure insecurity a constraint to investment. Instead, farmers reported that the greatest limitations were related to constrained access to inputs and credit, high input prices and limited availability of draft power (Muchetu 2019: 50).

Scoones et al. (2011) note that farms that had been secured by elites through political manipulation had markedly reduced output compared to before the reform. Corruption in allocating land to ZANU-PF party officials and urban elites retarded investment and productivity on large holdings near Harare, Zimbabwe's capital. Marongwe (2011) studied the productivity and investment effects of the FTLR program in Goromonzi District near Harare. He found that much of the land was unfarmed and that levels of production and productivity were significantly lower than in the pre-reform period. Marongwe (2011) attributed these results to the capture of the land allocation institutions by ruling party members and state security representatives. Most beneficiaries belonged to the governing or local elite and lacked the experience needed for managing commercial farming operations (Marongwe 2011).

Overall, the FTLRP did not enable Zimbabwe to maintain the same level of agricultural production that existed prior to the reform. But FTLRP smallholders seized on new commercial opportunities where finance and marketing conditions were right, particularly in the tobacco sector. Zimbabwe is among the top three tobacco producers in the world (Ngarava 2020). Before FTLRP in 2000, 98% of Zimbabwe's tobacco was produced by 2,000 large-scale, white-owned commercial farmers. By 2012, FTLRP beneficiaries dominated the tobacco industry, so that only 21% of tobacco was produced on large farms and 26% on medium-sized farms. Fifty-three percent of tobacco was produced on small-scale farms (Ngarava 2020). The tobacco industry provided farmers finance and other services not available to them for farming other commodities. Scoones et al. (2011: 972) point out that a radical reform such as FTLRP will be disruptive to production initially, but the larger concerns are the length of the transition to a new, potentially more inclusive agricultural sector, and what form the sector will take. Scoones et al. (2011) already found evidence of considerable household investment among 400 FTLRP settlers they studied in Masvingo District from 2008 to 2009. The settlers invested an average of $2,000 per family [extrapolated to US$91 million across all new resettlements] based on calculations of costs of labor and material for land clearance, housing and other buildings, and cattle and other farm equipment. Even without formal title or leasehold tenure, settlers had invested at scale, casting doubt on arguments that formalization of tenure regimes is necessary to catalyze investment. (Scoones et al. 2011: 983).

A 2014 study by Scoones found that off-farm work is important for meeting overall livelihood requirements of farming families in both the FTLRP resettlement areas and in the communal areas, recalling that most FTLRP beneficiaries are former residents of overcrowded communal areas. The new resettlement areas and the communal areas both have diversified economies, but the resettlement areas are more self-reliant, relying less on remittance flows and labor migration, and instead generating employment not only from the farms, but also from work as entrepreneurs, service providers, traders, and other businesses (Scoones 2014).

Moyo (2011) acknowledges the lower levels of investment and productivity of some crops by African farmers post-resettlement but argues that the greater achievement of the reform was restructuring of the agrarian economy. Moyo points out that FTLRP was largely successful in transferring most of Zimbabwe's white-owned commercial farming land to land reform beneficiaries in the form of small and medium-sized parcels. Due to land redistribution, 13% of Zimbabwe's farmland is held by medium-scale farmers, and over 70% is held by small farm producers (in the Communal Areas), in A1 areas and in informal settlements (Moyo 2011: 499). Moyo (2011: 499) concludes that FTLRP succeeded in bringing about "a net transfer of wealth and power from a racial minority of landed persons to mostly landless and land-poor classes."

Aliber and Cousins (2013) note that the large-scale commercial farming model that dominated Zimbabwe's agricultural sector prior to the FTLRP relied on state input and marketing subsidies and guaranteed price supports to maintain profitability. It is also important to underscore that the principal aim of FTLRP was restorative justice, that is, returning land to the indigenous African community forcibly displaced by white British settlers beginning in the 19th century. The FTLRP succeeded in distributing massive amounts of land to large numbers of rural African beneficiaries. Many of the post-reform farming enterprises were not commercially oriented by design (A1), and the commercial units (A2) were smaller and not capitalized to a degree comparable to the larger and previously white-owned commercial holdings.

Women's outcomes

Land invasions led by men were a principal means of securing land allocations in the critical early stages of the FTLRP, marginalizing women who were unable to take land by force (Matondi 2012: 206). When plots secured through land invasion were ultimately registered, inclusion of married women as joint holders of land alongside their husbands was rare. Mutopo (2011) found that women—mostly single, widowed, or divorced—comprised 18% of individuals who received A1 resettlement offer letters or A2 resettlement leases in Mwenezi District. Moyo et al. (2009: 26) found that 19% of women acquired land in their own right, with 15% securing land on A2 schemes. Scoones et al. (2010) found that in Masvingo District female-headed households were 8% of A2 beneficiaries and 13% held A1 plots.

Mutopo's research in Mwenezi District focuses on strategies used by married women to gain control of land held in the name of their husbands. She explores how women use "cultural negotiations" to gain control over discrete parcels of land within the household's FTLRP allocation on Merrivale Farms. She describes a strategy used by women to access land involving negotiation and bargaining with the state, family and traditional authorities, a strategy she considers preferable to the Western liberal approach that emphasizes individual human rights to land (Mutopo 2011: 1043). Mutopo concludes that while relying on husbands for access to land brought some degree of insecurity, women's livelihood strategies based on cross-border trade in South African urban markets were made possible largely due to the greater amount of land made available to them once resettled on FTLRP land (Mutopo 2011).

Addison (2019) concludes that women were empowered by the FTLRP reforms, though the gains were fragile. Women who are legally designated household heads and widows are in more secure positions than married women granted tseu (plots of land designated for use by wives by their husbands) Addison (2019: 104). Zikhali (2010) notes that women household heads might have been less likely to be recruited or less likely to apply given their greater reliance on non-farm income sources that conflicted with FTLRP's focus on agricultural households. She also argues that customary norms reserving land for men was a barrier to women's land access. Matondi (2012: 185) describes FTLRP as "a revolution without change in women's land rights." Matondi states:

> FTLRP was oblivious to the socio-cultural contexts within which women's access to, ownership of and control of land are mediated, interpreted and negotiated. People were allocated land in a policy vacuum because…the government had no time to develop an elaborate land policy that could have taken women's concerns on board. It was precisely because of this land policy vacuum that men took advantage of land occupations.
>
> (Matondi 2012: 206)

Contextual factors affecting outcomes

There were two contextual factors, largely overlooked or unanticipated, that had significant effects on outcomes.

Land invasions

The first was the land invasions that emerged seemingly spontaneously when beneficiaries living in nearby communal areas, informal settlements, and elsewhere took notice as FTLRP began to acquire white-owned commercial farming land (without compensation) on land that the government placed under state ownership. Led by war veterans and traditional authorities, the invasions resulted in rapid occupation of land by thousands of aspiring farming families. The government was unprepared to provide planning, technical, or administrative services, nor official land tenure security. There was considerable confusion and chaos during the first few growing seasons, as new residents strived to set up farming operations and non-farm enterprises. The occupations represented direct popular action to secure land and other resources denied them and their ancestors as a result of colonial conquest.

Despite a lack of international recognition and sanctions on the part of some Western countries, unclear tenure arrangements, and weak local government, many settlers strived and succeeded in building new communities based on agriculture and non-farm enterprise. Not all settlers succeeded. Still, the new settlements are not a replication of communal areas, "or scaled-down versions" of the commercial sector. Rather, they "are very different places with new people with new production systems engaging in new markets – all with new opportunities and challenges" (Scoones et al. 2011: 988).

Legitimacy of traditional authorities

Second, tenure insecurity was not perceived as an impediment to establishment of farms and homes, and beneficiaries felt confident in their ability to occupy and invest in land. In many areas, traditional authorities, with growing social and political legitimacy in the eyes of the members of resettlement communities and the state, provided informal land administration services and helped solve local land disputes. Still, FTLRP land is state land, and the task of delivering formal land rights currently remains in the hands of the state. Official policy toward tenure in the FTLRP areas is wedded to current systems of permits for A1 farms and 99-year leases for A2 holdings. FTLRP beneficiaries surely reflect a diversity of opinion, but many may find title deeds a preferable option, or even some form of customary tenure.

The Movement for Democratic Change (MDC), elected to power in 2018, may be in a position to build political consensus around a pluralistic approach to national tenure policy, accommodating a variety of tenure models—public, private, and customary—each suited to the needs and circumstances of different sections of the national community, while preserving Zimbabwe's achievements of transferring millions of hectares of farming land to thousands of African farmers. Scoones et al. (2011) argue that no one form of tenure provides a one-size fits all solution to tenure security. Rather, existing legislation provides for a variety of tenure types, "including freehold title, regulated leases, permits and communal tenure" (Scoones et al. 2011: 989).

Revised CMO (program theory) for Zimbabwe's land redistribution program

Table 7.1 depicts our revised CMO for Zimbabwe's land redistribution program. The FTLRP was successful in catalyzing redistribution of extensive areas of historically white-owned commercial agricultural land to land-short farmers from overcrowded communal areas (Moyo 2011: 941). Reform beneficiaries, impatient with the slow pace of land acquisition and redistribution since independence in 1980 and empowered by the government's decision to acquire land without compensating white commercial farmers, mounted land invasions that enabled rapid settlement of confiscated farms. Newly acquired commercial land, formally held under title deeds, was converted to state ownership. Beneficiary farmers were to receive permits or leases issued by the government. Lack of administrative capacity impeded the issuance of permits to A1 smallholder farmers and A2 medium to large-scale farmers. Still, newly settled farmers for the most part did not perceive tenure insecurity to be a constraint to investing in their new and larger holdings. Traditional authorities established themselves in FTLRP resettlement areas, claiming historical rights to land from which their ancestors had been displaced during the colonial era. In many settings, traditional authorities garnered greater legitimacy over land matters than local entities established by the central government (VIDCOs and RDCs). The roles of traditional authorities in rural governance were enshrined in the 2013 constitution and the 1998 Traditional Authorities Act. Nonetheless, the limits and scope of their powers over land matters are uncertain in light of the continuing interest of state authorities in land tenure matters in FTLRP areas. Commercial and state-managed credit programs for the most part failed to materialize, as did input supply programs (Muchetu 2019: 50). Thousands of small and medium-sized FTLRP farmers participated in tobacco production under contract farming arrangements financed by tobacco companies and banks, demonstrating a strong investment response where finance, marketing, and technical assistance are available. The most significant outcome of FTLRP was a restructured agrarian economy and rural landscape, settled by African farmers, who have used their own resources, including income from non-farm enterprises and wage remittances, to establish new farms, livelihoods, and communities. Tenure in most resettlement areas is generally secure and socially inclusive. Steps are needed to ensure that post-redistribution tenure policies don't enable a reconcentration of land, and ensure women have rights to land on par with men.

Table 7.2 Revised CMO for Zimbabwe's land redistribution program

1.0 Key contextual factors identified in the studies as affecting program outcomes (see Annex 6)

- Acquisition of white-owned commercial farms without compensation and organized land invasions on the part of intended reform beneficiaries from communal areas accelerated resettlement on FTLRP farms (Matondi 2012; Scoones et al. 2011)
- Central and local government officials and judiciary left unprepared to deliver tenure documentation to settlers. Private deeded commercial land was gazetted as state land. Many settlers received offer letters to their plots, but these were interim documents, provided in advance of eventual issuance of permits (for A1 farms) and leases (for A2 farms). Poor rates of delivery of formal documents, apart from issuance of leases to political elites and party officials living near Harare (Matondi 2012; Nyoni 2016; Scoones et al. 2011)
- Subsidized inputs for FTLRP beneficiaries are inconsistently available and in lesser amounts than pre-reform subsidies for white commercial farmers (Muchetu 2019)
- Low public investment in agriculture (credit, inputs, draft power in short supply), though evidence of significant investment by A1 and A2 farmers from their own savings and employment from non-farm enterprises (Muchetu 2019; Scoones et al. 2011). Tobacco sector came to be dominated by black smallholders in response to advance credit and prompt cash payment provided by buyers (Ngarava 2020)
- A restructured agrarian economy and rural landscape provides foundations for longer term and more inclusive rural growth (Moyo 2011; Scoones et al. 2011)
- Gender bias (against women) in allotments (Matondi 2012; Zikhali 2010), though some wives able to negotiate agreement with husbands for larger allotments for market production under their control (Mutopo 2011)

2.0 Observed outcome pathways following FTLRP

3a. Redistributive effect	**3b. Investment effect**	**3c. Market effect**
4a. Mechanism	**4b. Mechanism**	**4c. Mechanism**
Resource: farmland is available to farmers from communal areas in some cases; in other cases, allotments are given to urban or political elites *Response*: In rural areas: beneficiaries perceive that land and input supports will increase farm-based incomes Near urban areas: urban elites see land as a speculative investment or for retirement	*Resource*: ambiguous formal rights for landholders *Response*: in many cases, familiar customary tenure norms are applied in ad hoc fashion, and in other settings customary authorities step in to fill the gap in land tenure services, providing largely effective land administration in A1 areas *Response*: generally speaking, tenure insecurity not cited as constraint to investment, though poor formal credit supply is considered a problem by some, especially A2 farmers	*Resource*: ambiguous rights for landholders stymied credit access (especially problematic for A2 farmers) and subsidized inputs provided sporadically *Response*: landholders seeking to shift from subsistence to commercial production find greatest success in contract tobacco farming, where inputs are provided by buyers and payments are quick

(*Continued*)

Table 7.2 (Continued)

5a. Intermediate outcome In rural areas: landholders make soil quality improvements but sporadic availability of inputs and credit limits productivity. Evidence, though, of considerable on investment in fencing, farm equipment and housing (Scoones et al. 2011) Near urban areas: elites put only limited areas of farms under cultivation. Overall productivity is low	**5b-1. Intermediate outcome 1** Landholders perceive tenure is secure and many invest their own savings and off-farm income in farming. Some farmers increase reliance on off-farm incomes to meet livelihood needs and subsidize farming operations **5b-2. Intermediate outcome 2** Landholders, because of limited cash and credit, don't invest in soil improvements and fertilizer and other inputs and levels of productivity are low. Large number of smallholders enter into contract tobacco farming arrangements where inputs are provided by buyers	**5c-1. Intermediate outcome 2** In rural areas: landholders make soil quality improvement investments but the high cost and sporadic availability of inputs and credit limits productivity increases Near urban areas: much redistributed land is not farmed or only small portions of it are put into production
6a. Final outcomes Seventy-five percent of white-owned commercial farmland distributed mainly to African smallholders. More people have access to larger holdings than available in communal areas, on more fertile land. However, allotments near urban areas, especially Harare, are disproportionately given to urban elites/political leaders In rural areas, agricultural production increases relative to communal areas	**6b. Final outcome** Despite poor delivery by state institutions of formal tenure (A1 permits and 99-year leases to A2 farmers) land tenure is perceived to be secure for most LTLRP beneficiaries. Divisions on national land tenure policy persist	**6c. Final outcome** [in rural and near urban areas] Agricultural production declines relative to pre-reform production

References

Addison, L. (2019). The fragility of empowerment: changing gender relations in a Zimbabwean resettlement area. *Review of African Political Economy*, *46*(159), 101–116. https://doi.org/10.1080/03056244.2019.1610939.

Aliber, M., & Cousins, B. (2013). Livelihoods after land reform in South Africa. *Journal of Agrarian Change*, *13*(1), 140–165. https://doi.org/10.1111/joac.12012.

Hull, S., Babalola, K., & Whittal, J. (2019). Theories of land reform and their impact on land reform success in Southern Africa. *Land*, *8*(11), Article 11. https://doi.org/10.3390/land8110172.

Marongwe, N. (2011). Who was allocated fast track land, and what did they do with it? Selection of A2 farmers in Goromonzi District, Zimbabwe and its impacts on agricultural production. *The Journal of Peasant Studies*, *38*(5), 1069–1092. https://doi.org/10.1080/03066150.2011.636483.

Matondi, P. B. (2012). *Zimbabwe's Fast Track Land Reform*. Bloomsbury Publishing.

Mkodzongi, G. (2016). 'I am a paramount chief, this land belongs to my ancestors': the reconfiguration of rural authority after Zimbabwe's land reforms. *Review of African Political Economy*, *43*(sup1), 99–114. https://doi.org/10.1080/03056244.2015.1085376.

Mkodzongi, G., & Lawrence, P. (2019). The fast-track land reform and agrarian change in Zimbabwe. *Review of African Political Economy*, *46*(159), 1–13. https://doi.org/10.1080/03056244.2019.1622210.

Mlambo, A. S. (2014). *A History of Zimbabwe*. Cambridge University Press.

Moyo, S., Chambati, W., Murisa, T., Siziba, Dangwa, C., Mujeyi, K., & Nyoni, N. (2009). Fast track land reform baseline survey in Zimbabwe: trends and tendencies, 2005/06. Harare: African Institute of Agrarian Studies.

Moyo, S. (2011). Three decades of agrarian reform in Zimbabwe. *The Journal of Peasant Studies*, *38*(3), 493–531. https://doi.org/10.1080/03066150.2011.583642.

Muchetu, R. G. (2019). Family farms and the markets: examining the level of market-oriented production 15 years after the Zimbabwe Fast Track Land Reform programme. *Review of African Political Economy*, *46*(159), 33–54. https://doi.org/10.1080/03056244.2019.1609919.

Mutopo, P. (2011). Women's struggles to access and control land and livelihoods after fast track land reform in Mwenezi District, Zimbabwe. *The Journal of Peasant Studies*, *38*(5), 1021–1046. https://doi.org/10.1080/03066150.2011.635787.

Ngarava, S. (2020). Impact of the Fast Track Land Reform Programme (FTLRP) on agricultural production: a tobacco success story in Zimbabwe? *Land Use Policy*, *99*, 105000. https://doi.org/10.1016/j.landusepol.2020.105000.

Nyoni, J. (2016). *Land Tenure and Land Marketability: Policy Options and Recommendations* (USAID Economic Research and Analysis–Zimbabwe (SERA) Program). Nathan Associates.

Scoones, I. (2014, August 4). Comparing communal areas and new resettlements in Zimbabwe V: farm employment, off-farm income earning and livelihood diversification. *Zimbabweland*. https://zimbabweland.wordpress.com/2014/08/04/comparing-communal-areas-and-new-resettlements-in-zimbabwe-v-farm-employment-off-farm-income-earning-and-livelihood-diversification/.

Scoones, I., Marongwe, N., Mavedzenge, B., Murimbarimba, F., Mahenehene, J., & Sukume, C. (2010). *Zimbabwe's Land Reform: Myths and Realities.* Oxford: James Currey; Harere: Weaver Press; Johannesburg: Jacana.

Scoones, I., Marongwe, N., Mavedzenge, B., Murimbarimba, F., Mahenehene, J., & Sukume, C. (2011). Zimbabwe's land reform: challenging the myths. *The Journal of Peasant Studies*, *38*(5), 967–993. https://doi.org/10.1080/03066150.2011.622042.

Zikhali, P. (2010). Fast track land reform programme, tenure security and investments in soil conservation: micro-evidence from Mazowe District in Zimbabwe. *Natural Resources Forum*, *34*(2), 124–139. Scopus. https://doi.org/10.1111/j.1477-8947.2010.01298.x.

Zikhali, P., & Chilonda, P. (2012). Explaining productivity differences between beneficiaries of Zimbabwe's Fast Track Land Reform Programme and communal farmers. *Agrekon*, *51*(4), 144–166.

8 Synthesis of findings

Patterns in PDR/LRD program outcomes

Our realist synthesis of parcel demarcation and registration (PDR) programs in Benin, Ethiopia, and Rwanda and the land redistribution (LRD) program in Zimbabwe's revealed some striking differences across the cases, as well as some common themes in the outcomes of interest and the contextual factors associated with them.

The studies from Ethiopia and Rwanda support the argument that parcel demarcation and rights registration can enhance tenure security for significant segments of the population. The studies from Benin, however, tell a more nuanced story; intended beneficiaries showed limited interest in acquiring certificates, gains in tenure security were generally less apparent, and latent land conflicts erupted in areas with significant numbers of migrants and tenant farmers. Evidence from Zimbabwe's Fast Track Land Reform Programme (FTLRP) suggests that while some form of statutory title was to be granted LRD beneficiaries, often ad hoc customary tenure arrangements emerged as a rough and ready solution to the need to deliver rights on the ground. This was due to a combination of landholder preference, successful political maneuvering by traditional authorities, and the inability of state authorities to deliver a title option perceived by reform beneficiaries to be credible.

Evidence that PDR programs have positive outcomes for land conservation investments and agricultural productivity is strongest for Ethiopia and inconsistent for Benin and Rwanda. Zimbabwe's FTLRP for the most part showed enhanced agricultural production compared to communal areas, but lower production levels than white-owned commercial farms. We found evidence that levels of investment in soil conservation were higher in communal areas than on FTLRP farms. Since impacts on agricultural productivity are likely to take several years to materialize, we must remain cautious when drawing conclusions about both the impacts and the contextual factors affecting them until households included in the early evaluation are re-evaluated later.

The theorized credit effect did not materialize in Ethiopia (where use of land as collateral is prohibited) or in Benin, and it was not discussed in the

DOI: 10.4324/9781003365679-8

Zimbabwe LRD studies. A small percentage of landholders in Rwanda used land titles to obtain credit, but little of the borrowed money was spent on agricultural investments, going primarily to housing construction.

Although informal land sales are robust in Rwanda, it is not clear that they have increased since implementation of the LTR program. In Ethiopia and Rwanda, possession of land certificates is linked to increases in land rentals, a development that appears to be beneficial for female heads of household, typically divorcees or widows with limited access to male labor and draft power.

Social inclusion outcomes for women are generally positive for Rwanda and Ethiopia, albeit with some exceptions. The limited evidence from Benin suggests that social inclusion outcomes for women (in terms of enhanced tenure security) were weakly positive overall, though likely negative for many women (Giovarelli et al. 2015).

In Benin, the only case in which tenure security outcomes for migrants were studied, the limited evidence suggests that tenure security outcomes were negative for short-term migrants and positive for long-term migrants.

Future research themes for understanding contextual factors affecting PDR and LRD program outcomes

1 Relative functionality of customary and state land governance systems

Our review suggests that the relative functionality and legitimacy of customary and state land governance systems, as well as how they interact, is a key contextual factor likely to influence PDR and LRD program outcomes. Tenure security gains were more evident in Ethiopia and Rwanda, where customary land governance systems have experienced severe disruption and state support for certification and capacity to deliver land certificates is stronger than in Benin, where customary land governance systems retain widespread legitimacy. The low demand for land certificates in Benin, coupled with low incidence of land conflict in most areas and high confidence in local land governance institutions to resolve disputes, suggests that relatively few landholders perceive their tenure to be insecure. Nonetheless, there are some segments of rural society in Benin, notably women and long-term migrants, for whom existing rules and norms of the customary tenure system are less able to provide satisfactory levels of tenure security and who may benefit from tenure interventions that bolster their security without abandoning traditional authority for state control and greater tenure individualization.

Based on research in Senegal and Zambia, Honig (2017) develops the useful concept of "customary privilege." Households with higher privilege, measured by strong social links to local customary authorities, are more likely to perceive that customary tenure provides strong security and have low demand for titles. But households with lower privilege are more likely to have greater demand for state titles because their customary property rights are weaker (Honig 2017: 103). Titling (and certification) is typically understood as an intervention targeted narrowly at incentivizing investment by individual households in their farming enterprises when in fact it brings about major institutional changes, capable of radically transforming entrenched power structures within communities(Honig 2017: 104). Most post-intervention studies we examined focused mostly on the investment and productivity outcomes of interventions and rarely on the effects of individualization on the broadly inclusive character of customary regimes, despite inequalities in status within the social structure. Unless reforms protect the current beneficiaries of customary systems, while also promoting social inclusivity, they will likely meet with considerable resistance, or indifference (Miceli et al. 2001).

In Zimbabwe, poor delivery of A1 permits and A2 leases in resettlement areas, debates about the role of titles and land markets, and contestation over claims by traditional authorities to historic rights to FTLRP land combined to heighten perceptions of tenure insecurity among some beneficiaries of FTLRP allocations. Still, FTLRP succeeded in redressing to a considerable degree the historic injustices of colonial era land theft, displacement, and impoverishment, by transferring thousands of hectares of white-owned commercial farming land to small and medium-scale African farmers resettled from the communal areas. Ad hoc tenure arrangements, in some instances drawing on familiar customary tenure norms, proved adequate to most reform beneficiaries in resettlement areas (Scoones 2020). The state's inability to ensure adequate supplies of inputs and access to credit for beneficiaries of the A1 and A2 resettlement programs helped undermine the government's goals of maintaining pre-reform agricultural productivity levels. Resettlement beneficiaries, like small farmers across Africa, find it difficult to make a living from farming alone, and have diversified household livelihoods to include other livelihood activities alongside farming.

2 Disjuncture between complex overlapping customary rights and individual rights registration

A contextual factor that was particularly notable in the Benin case was the gap between the complexity and logic of customary tenure systems and PDR's focus on registering individual or household rights to land. As Boone (2019) points out, tenure interventions that simplify existing rights risk accelerating processes of disempowerment and land dispossession of more vulnerable members of society. This was evident in the studies by Lavigne Delville and Moalic (2019) and Yemadje et al. (2012, 2014) which reported that founding lineage and extended family heads sought to be listed as primary rights holders so that they could retain control over lands they had granted to others, control which was the source of both their wealth and political power. Disadvantaged were short-term migrants, women, and tenant farmers, groups which typically relied on negotiated agreements with more powerful community members and who were in a weak bargaining position when land rights were being recorded.

3 Parcel demarcation as a tool for stabilizing disrupted customary systems

In Rwanda and Ethiopia, where customary tenure systems appeared to be weakened by political and social conflicts and where holdings were already more individualized, rights registration may have had the salutary effect of reducing rather than aggravating conflicts over land. That said, the suggestion that in those countries certification provided an alternative, state-based system to customary tenure for assuring land rights would be overstated. An alternative hypothesis is that largely viable customary institutions and norms were disrupted by the Derg land reforms in Ethiopia, and that in parts of Rwanda widespread inter-communal mixing driven by internal migration upset the ethnic balance in many localities, contributing to severe land conflicts, providing an important antecedent to the genocide. The greatest benefits of the certification programs may have been the stabilization of the rural social systems in both countries, reducing land-based social conflict and social and economic uncertainty without abridging key principles of the customary systems. In Ethiopia, many tenets of pre-Derg customary tenure were enshrined in the national and regional state proclamations, including prohibitions on land sales, a key social protection in customary tenure systems across Sub-Saharan Africa. According to Todorovski and Potel (2019: 1), the post-conflict Rwandan government believed that orderly land rights administration was important to the maintenance of a sustainable peace because it enhances social equity and prevents conflicts.

4 Gender equality laws, social norms, and interventions supporting women's land rights

In all three PDR cases, state legal frameworks supported gender equality in land access and inheritance. These encompassed marriage and family law and/or succession law as well as land law, highlighting the importance of complementary reforms in sectors other than land governance in solidifying individual and joint ownership rights for all. Laws identified in the studies as important for positive social inclusion outcomes with respect to gender include those that specify equal inheritance rights for all children regardless of sex and inheritance rights for women in cases of divorce or widowhood. Also critical for positive gender inclusion outcomes are laws that grant women direct instead of derived rights. In Zimbabwe, the stated policy aims of FTLRP called for gender equity in choosing land reform beneficiaries, but inattention to gender considerations in program implementation, exacerbated by the disorder associated with the land invasions, contributed further to unequal outcomes. In the PDR cases, family and marriage laws sometimes contributed to negative social outcomes. Women in informal unions tended to benefit less than those in state-registered marriages in Ethiopia (Cloudburst 2016; Kumar and Quisumbing 2015) and Benin (Giovarelli et al. 2015), and in Benin, women in polygamous unions were at a distinct disadvantage since the state does not consider polygamous unions to be legally valid. In Rwanda, women in polygamous marriages who were not legally married appeared able to register the plots they used in their own names, with agreement of their husbands. They also registered their own children as heirs. Most studies emphasized the need for recognizing the land rights of all women regardless of their marriage status and whether they are in legal unions or not.

Persistent cultural norms that prevented women from actualizing their land rights attenuated the beneficial effects of legal frameworks supportive of gender equality in all three PDR cases. These norms hinder women from being included on certificates for jointly held land, especially in Benin and Rwanda. While reforms clarified women's rights to inherit and in some cases to own land independently, none appear to fundamentally disrupt the patrilineal social order on which customary tenure rests in most areas of those countries. Tenure security outcomes for women were more favorable in Rwanda and Ethiopia than in Benin. One likely factor is that Ethiopia and Rwanda's PDR programs had joint titling requirements for married couples, whereas in Benin, project guidelines encouraged joint titling for spouses, but did not require it. In Benin, efforts were made to incorporate awareness of the importance of registering secondary rights, but they were implemented late in the project and inconsistently across the targeted communes. Additionally, both Ethiopia and Rwanda integrated into their programming extensive land and inheritance rights education and awareness programs for women, as well as training in women's land rights for community leaders. However, as Bayisenge (2018) notes, gains women made in understanding their rights did not guarantee them a role in community leadership on land matters or within formal administrative services. In Benin, delayed implementation and limited geographical coverage hampered the effectiveness of women's land rights awareness programming.

Some scholars argue that the conditionality of women's (and men's) access to land on family and ultimately descent group rights is a valuable feature of customary tenure and should not be disturbed without considerable forethought (Djurfeldt 2020; Peters 2020). In this formulation, stronger individual rights are antithetical to achievement of the customary system's greater social benefits. Henrysson and Joireman (2009: 41) point out that customary legal decisions tend to be based on compromise, in the interest of maintaining the social relationships that bind the social system together (Henrysson and Joireman 2009). Peters (2020: 47) argues that to sweep away "a practice of land allocation considered discriminatory will disrupt the basis for kinship-based practices of cooperation and interdependence on which rural life (and much of urban life) depend." The effects of reforms granting women state-backed equality of ownership on customary tenure's "web of social relations" (Peters 2021: 47) that support a population with few other means of economic security merits serious attention in future research.

5 Disjuncture between rural livelihood strategies and policy focus on agriculture

In Zimbabwe, the FTLRP reconfigured the agrarian structure (Aliber and Cousins 2013) without transforming most small-scale holdings into commercial enterprises. Rather, the program allowed rural homesteads to become bases for diversified income earning strategies on holdings larger than previously available to them in the communal areas. None of the studies for Ethiopia, Rwanda, or Benin explored how agricultural productivity or investment outcomes of PDR programs affected household participation in non-farm livelihood activities or how household non-farm activities affected those outcomes. However, the results from Zimbabwe, as well as studies of SSA livelihood strategies (Gassner et al. 2019) suggest that reliance by rural families on non-farm income can have significant effects, both positive and negative, on the degree to which households invest labor and money in farming, including when perceptions of tenure security are high. Future evaluations of the agricultural investment effects of interventions intended to strengthen tenure security should more fully consider the significance of non-farm income and employment on farm-level investment decisions.

6 Medium-sized farms as emerging drivers of landholding inequality

Jayne et al. (2014) and other recent studies by Jayne and his colleagues (e.g., Jayne et al. 2019) raise concerns that the expanding middle-sized farm sectors (consisting of holdings between 5 and 50 hectares) in several countries, including Zambia, Tanzania, and parts of Ghana and Nigeria, will decrease the land available for smallholders in the future, resulting in greater structural inequality. Each of these countries has relatively abundant amounts of unsettled land, providing space where the middle-scale farming sector can expand, and for now avoiding direct competition with areas of dense agricultural land use predominantly held by smallholders under customary tenure. This contrasts with Rwanda and Ethiopia, two of the countries we studied, which are emerging from acute political conflict driven in part due to severe land shortage. Jayne and his colleagues draw attention to the need for land policies that would ensure availability of land for smallholder expansion in the future. Expansion of a medium- to large-farm sector may take place in areas under state tenure that could also, or instead, be used to accommodate the expansion needs of a smallholder sector governed under customary tenure principles.

References

Aliber, M., & Cousins, B. (2013). Livelihoods after land reform in South Africa. *Journal of Agrarian Change*, *13*(1), 140–165. https://doi.org/10.1111/joac.12012.

Bayisenge, J. (2018). From male to joint land ownership: women's experiences of the land tenure reform programme in Rwanda. *Journal of Agrarian Change*, *18*(3), 588–605. https://doi.org/10.1111/joac.12257.

Boone, C. (2019). Legal empowerment of the poor through property rights reform: tensions and trade-offs of land registration and titling in Sub-Saharan Africa. *The Journal of Development Studies*, *55*(3), 384–400. https://doi.org/10.1080/00220388.2018.1451633.

Cloudburst Group. (2016). *USAID Land Tenure ELTAP-ELAP Impact Evaluations Endline Report*. https://www.land-links.org/wp-content/uploads/2016/09/USAID_Land_Tenure_ELTAP-ELAP_Impact_Evaluations_Endline_Report.pdf.

Djurfeldt, A. (2020). Gendered land rights, legal reform and social norms in the context of land fragmentation a review of the literature for Kenya, Rwanda and Uganda. *Land Use Policy*, *90*, 104305. https://doi.org/10.1016/j.landusepol.2019.104305.

Gassner, A., Harris, D., Mausch, K., Terheggen, A., Lopes, C., Finlayson, R., & Dobie, P. (2019). Poverty eradication and food security through agriculture in Africa: rethinking objectives and entry points. *Outlook on Agriculture*, *48*(4), 309–315. https://doi.org/10.1177/0030727019888513.

Giovarelli, R., Hannay, L., Scalise, E., Richardson, A., Seitz, v., & Gaynor, R. (2015). *Gender and Land: Good Practices and Lessons Learned from Four Millennium Challenge Corporation Compact Funded Land Projects. Synthesis Report and Case studies: Benin, Lesoto, Mali, and Namibia*. Landesa and Center for Women's Land Rights. https://resourceequity.org/record/2734-gender-and-land-good-practices-and-lessons-learned-from-four-millennium-challenge-corporation-compact-funded-land-projects/.

Henrysson, E., & Joireman, S. F. (2009). On the edge of the law: women's property rights and dispute resolution in Kisii, Kenya. *Law & Society Review*, *43*(1), 39–60. https://doi.org/10.1111/j.1540-5893.2009.00366.x.

Honig, L. (2017). Selecting the state or choosing the chief? The political determinants of smallholder land titling. *World Development*, *100*, 94–107. https://doi.org/10.1016/j.worlddev.2017.07.028.

Jayne, T. S., Chapoto, A., Sitko, N., Nkonde, C., Muyanga, M., & Chamberlin, J. (2014). Is the scramble for land in Africa foreclosing a smallholder agricultural expansion strategy? *Journal of International Affairs*, *67*(2), 35–53.

Jayne, T. S., Muyanga, M., Wineman, A., Ghebru, H., Stevens, C., Stickler, M., Chapoto, A., Anseeuw, W., Westhuizen, D., & Nyange, D. (2019). Are medium-scale farms driving agricultural transformation in sub-Saharan Africa? *Agricultural Economics*, *50*(S1), 75–95. https://doi.org/10.1111/agec.12535.

Kumar, N., & Quisumbing, A. R. (2015). Policy reform toward gender equality in ethiopia: little by little the egg begins to walk. *World Development*, *67*(C), 406–423.

Lavigne Delville, P., & Moalic, A.-C. (2019). Territorialities, spatial inequalities and the formalization of land rights in Central Benin. *Africa*, *89*(2), 329–352. https://doi.org/10.1017/S0001972019000111.

Miceli, T. J., Sirmans, C. F., & Kieyah, J. (2001). The demand for land title registration: theory with evidence from Kenya. *American Law and Economics Review*, *3*, 275–287.

Peters, P. E. (2020). The significance of descent-based 'customary' land management for land reform and agricultural futures in Africa. In C. M. O. Ochieng (Ed.), *Rethinking Land Reform in Africa New Ideas, Opportunities and Challenges* (pp. 70–83). African Development Bank. https://www.afdb.org/en/initiatives-partnerships/african-natural-resources-centre/publications/rethinking-land-reform-africa-new-ideas-opportunities-and-challenges.

Peters, P. E. (2021). Kinship. In A. H. Akram-Lodhi, K. Dietz, B. Engels, & B. McKay (Eds.), *Handbook of Critical Agrarian Studies* (pp. 139–149). Edward Elgar Publishing. https://www.elgaronline.com/display/edcoll/9781788972451/9781788972451.00024.xml.

Scoones, I. (2020, August 27). Still debating land tenure reform in Zimbabwe. *Zimbabweland Blog*. https://zimbabweland.wordpress.com/2020/08/27/still-debating-land-tenure-reform-in-zimbabwe/.

Todorovski, D., & Potel, J. (2019). Exploring the nexus between displacement and land administration: the case of Rwanda. *Land*, *8*(4). https://doi.org/10.3390/land8040055.

Yemadje, R. H., Crane, T. A., Mongbo, R. L., Saïdou, A., Azontonde, H. A., Kossou, D. K., & Kuyper, T. W. (2014). Revisiting land reform: land rights, access, and soil fertility management on the Adja Plateau in Benin. *International Journal of Agricultural Sustainability*, *12*(3), 355–369. https://doi.org/10.1080/14735903.2014.909645.

Yemadje, R. H., Crane, T. A., Vissoh, P. V., Mongbo, R. L., Richards, P., Kossou, D. K., & Kuyper, T. W. (2012). The political ecology of land management in the oil palm based cropping system on the Adja plateau in Benin. *NJAS: Wageningen Journal of Life Sciences*, *60–63*(1), 91–99. https://doi.org/10.1016/j.njas.2012.06.007.

9 Conclusions

The principles underlying customary tenure—collective ownership, inclusion as a social right but with status differentiated by gender and other social factors—are among the most consequential contextual factors affecting the outcomes of land reform interventions. Some reforms have sought to reorder and disrupt these principles, in the name of promoting agricultural productivity, gender equity, and accountable governance. Customary systems have proven resistant to many—though not all—of these interventions, not because these systems are inherently regressive or backwards, but because they continue to serve the interests of large sections of their members. Most customary rights holders, both men and women, are poor, and in the face of uncertain and modest economic prospects in the larger economy find a measure of social and economic security in rural place-based social systems that provide, through customary tenure, access to land—a foundation of cultural identity, community solidarity, and livelihood opportunity—as a social right (Peters 2021; Ubink 2011).

Customary land rights, then, provide more than access to land; they are markers of membership in rural social systems that provide a multitude of benefits. The narrow focus on certification as necessary to catalyze investment in agriculture is more likely relevant in settings where the actual problem is general social instability of which tenure insecurity is but one expression. The experiences of Ethiopia and Rwanda have demonstrated that certification can contribute to social stabilization. But once social stabilization is achieved, investment in agriculture will be influenced by factors other than tenure security.

Programs should prepare mitigation strategies to confront the potential risk of strengthening customary privilege, and PDR programs should tailor land tenure interventions in ways that deliver tenure security to groups disadvantaged under existing customary systems (Honig et al. 2017: 104). In Zimbabwe, resettlement program officials might have made more constructive use of familiar forms of customary tenure in organizing settlement of reform beneficiaries on newly acquired commercial farms from the outset. After all, most beneficiaries originated from communal areas, where customary tenure norms prevailed. In fact, only a relatively small percentage of settlers reported

DOI: 10.4324/9781003365679-9

serious problems with tenure security, suggesting perhaps that beneficiaries sorted out allocations and resolved disputes relying on socially familiar customary norms. The political legitimacy and legal authority of customary chiefs in the FTLRP areas grew steadily from the outset of the program in 2000, especially in relation to local civic authorities. Mkodzongi (2016) found that as FTLRP progressed, chiefs took on many land administration tasks, including allocating land, though these powers had been formally invested in Rural Development Councils, bodies created by the central government. As we noted in Chapter 2, the FTLRP resettlement area became an arena for a real-time struggle between neotraditional and statist land tenure reform interests (Boone 2007). Understanding the issue in those terms can help policymakers, local leaders, and land users understand what's ultimately at stake.

Understanding the rural household economy in context

Most studies in the economics literature evaluating the effects of land rights formalization focus nearly exclusively on agricultural investment and productivity outcomes. Under customary tenure systems, land rights are not only forms of natural and economic capital but also serve as social, cultural, and political capital. Holding customary rights enables household members to participate in a variety of social, economic, and civic activities apart from farming, serving the general welfare of family members and contributing to long-term community stability (Butler 2021). Household labor and savings may be directed to a variety of non-farm activities, including wage employment, informal trading, and small businesses that may or may not have links to agricultural value chains. The growth of non-farm rural employment has been shown to generate financial capital for investment in farming (Ethiopia and Zimbabwe). On the other hand, the high opportunity costs of agricultural labor may lead some households to withdraw from farming altogether (Lesotho) or adopt labor-saving agricultural activities such as tree-planting (Ethiopia), while maintaining their commitment to the rural social system and the diverse non-agricultural benefits it provides (Murray 1981). When contemplating agricultural reforms, policymakers and program designers are more likely than not to encounter largely stable customary systems, providing a secure base for participation in the rural social system. Care should be taken not to disturb the rural social system in pursuit of misplaced efforts to correct a tenure problem that doesn't exist (Benin).

Socially inclusive and differentiated: mastering the paradoxes of customary tenure

Social inequalities between men and women and elites and non-elites within rural social systems can be significant, and may diminish the economic prospects, social wellbeing, and agency of women, children, and minority and migrant communities and non-elite men. Designers of tenure reforms need to be more aware of the significant social and economic benefits afforded by customary tenure, including a broad social inclusiveness of men and women based on descent group or lineage membership, while framing interventions not as incremental efforts at system replacement, but with the aim of closing the gap between those with high and low privilege *within* the system (Honig 2017: 104). State-led reforms promoting joint registration of land rights in Ethiopia and Rwanda have succeeded in reducing the privilege gap between husbands and wives by clarifying and strengthening land and inheritance rights of married women. But reforms also can come from within customary systems. Customary leaders in Lesotho changed customary inheritance rules to establish that widows and not eldest sons inherit agricultural land (Lawry 1993). Arguably, this had the effect of assuring widows of their status in the rural social system as rightful landholders, consolidating their support for the system. Traditional leaders in Namibia engaged in a customary law-making exercise that prohibited "land grabbing" and "widow dispossession," aligning customary norms with constitutional principles (Ubink 2011: 2). Reducing customary privilege will require careful and patient engagement with community members, with interventions based as much as possible on consensus, backed up by education and training programs, and strengthened by state and traditional enforcement mechanisms, to help women and other marginalized groups defend their new ownership, use, and inheritance rights (Fitzpatrick 2006).

Certification as an instrument of land tenure reform policy

Stakeholders approach tenure reforms with different assumptions about their goals and purposes. Donors and some government officials tend to promote the individualization, agricultural investment, and enabled land market goals as the most important, while others, including community members and advocacy groups, may focus on the stabilization, social inclusion, and gender equity elements of the program (the liberal pluralism discourses cited by Ayano 2018). Each of the reform efforts we have examined is riven by differences among proponents and detractors about their ultimate aims. Certification, then, emerges as a canvas on which different interests paint their own visions of what just, inclusive, secure, and productive tenure arrangements look like.

The purposes served by certification are not always given the explicit attention they warrant. Certification is an instrument of land tenure policy goals, and not merely an end. It is well known that certification can simultaneously have negative and positive effects on different sectors of communities. Drawing on Ayano (2018) we discern three policy reform contexts in SSA where certification is used, in each context serving different purposes and yielding quite different results.

- Certification is used to validate existing customary tenure arrangements, where these have been called into question due to internal social and economic changes or political shocks. Deininger et al. (2017: 3) cite examples of local demand for even informal affirmation of existing customary rights in Madagascar, Mali, Benin, Burkina Faso, and Rwanda. Ayano (2018: 2) characterizes policymakers' commitment to legally recording customary interests in land as an expression of "pluralist formalization" that acknowledges a great variety of local land customs and traditional practices. Knight (2010) draws attention to largely successful efforts in Botswana, Tanzania, and Mozambique to accord statutory recognition of customary tenure, at a legal status equivalent to co-existing private and public tenure.
- Certification may also be used as a means (or a resource in CMO parlance) to introduce selected, targeted new principles into customary systems. Certification may act to preserve many of the core principles of the systems or may set the customary system on a pathway toward fundamental change by, for instance, formally accommodating market principles. Ayano (2018: 8) speaks of a form of pluralism that combines two competing visions. One is "liberal pluralism," a globalist vision of multiculturalism, embracing progressive notions of rights, such as the rights of indigenous peoples and women, self-determination, and shared development. Examples would include joint certification of ownership by husband and wife as in Ethiopia and Rwanda, with the specific aim of clarifying the right of widows to inherit. The other embraces a vision of land as primarily an economic asset rather than a cultural or political institution, with individualized ownership and land markets catalyzing investment and higher farm productivity (Ayano 2018: 8).
- By valuing land as primarily an economic asset, certification is more likely to be used as an instrument of tenure reforms that aim to decisively transform core customary tenure system principles, or rights of access based on social status in descent groups, to state-issued marketable titles. Ayano (2018) cites evidence that certification has extended the power of the state and the wealthier in rural Ethiopian society without contributing to any progressive outcomes. Given persistent power imbalances, certification may prove in the longer-term a legal and political framework for the wider introduction of rural land markets, to the detriment of rural customary communities.

Development theory toward customary tenure remains stubbornly grounded in the assumption that it constitutes a severe institutional barrier to smallholder agricultural modernization. Uncritical acceptance of this assumption continues to lead policymakers to ill-starred investments in reforms intended to promote agricultural investment while failing to take account of the larger social, economic, and cultural benefits afforded by an existing, largely inclusive customary tenure system that provides secure access by poor people to land as a social right. More problematically, the continued insistence of policymakers and development practitioners on viewing customary tenure systems as backward and insecure has increased insecurity for smallholders and enabled internal and external groups to appropriate land. Without proper protection from these trends, customary landholders risk losing their land.

This realist synthesis has revealed how beneficiaries, in important instances, have used the resources provided by certification to build more socially inclusive and secure customary systems, and not to discard them.

Land tenure reform policy and the persistent failure to account for context

Tenure reform practitioners, particularly international advisors, tend not to take adequate account of the contemporary contexts within which reforms, narrowly intended to boost agricultural productivity, are introduced. The studies for Ethiopia and Rwanda show that, when introduced into contexts where social systems and customary arrangements have been disrupted by political conflict and social dislocation, PDR interventions reestablished and clarified rights, helping to stabilize local social systems and economies. This was arguably the principal goal of national politicians and planners, and both national programs largely succeeded in achieving these political goals. Modest gains to productivity and investment followed for some members of the landholding community, along with a host of other unenumerated social and economic benefits associated with post-conflict stabilization. However, in Benin, which did not have a prior history of massive social disruption, the intervention catalyzed latent land conflicts between founding lineages and migrants in a frontier area and between landlords and tenant farmers in a densely populated area. In Zimbabwe, the beneficiaries of land redistribution gained access to larger landholdings on which to potentially generate better, more stable incomes. Traditional authorities provided ad hoc tenure arrangements where state land rights certification and leasing programs faltered or failed to materialize; this was an example of the enduring legitimacy of customary tenure principles (Mkodzongi 2016; Scoones et al. 2011).

What explains the persistent failure to understand context before launching land reforms based on certification and individualization that risk corroding the institutional arrangements on which a broadly inclusive customary tenure system rests? The theoretical underpinnings of land tenure reform have for the most part been grounded in neoclassical economics. Surely then, one answer lies in the invisibility to Western-educated reform scholars, planners, and advocates of the social foundations and purposes of traditional institutions. The narrow focus on agricultural investment and productivity, and the lack of alternatives to individual title in the neoclassical policy property rights theory toolkit is a related explanation. This may also explain that even when investment and productivity responses in Sub-Saharan Africa to certification are weak (Lawry et al. 2017), scholars will hold onto their belief in the inherent insecurity of customary tenure, and search for explanations in other economistically familiar explanations. One common example attributes failure to weak investment in administrative and staffing capacity (Ayano 2018), without considering the possibility that national budget managers may observe that neotraditional arrangements continue to deliver tenure security quite adequately and prefer to prioritize building staff capacity in other sectors struggling to manage what they consider more urgent problems. It is not unusual for landholders, once they receive certificates, to fail to register transfers of ownership with local land offices following sales or inheritances (as found in Rwanda by Ali et al. 2019). A familiar explanation offered for failure to register transfers is that the costs of registering land transfers are too high, and if costs were lowered the rates of registration would increase. Here, an alternative explanation might be that most transactions of cropland are carried out within the descent group, which provides relatively strong assurance of tenure security, at almost no financial cost.

The realist synthesis method is designed to help program evaluators understand how relevant contextual factors influence intervention outcomes. The application of realist synthesis to the study of land tenure reform interventions has, we believe, provided evidence not only of the value of the method, but also of the vital of importance of investing in understanding context before launching major tenure reform interventions. But to properly understand context, the near exclusive authority that donors (especially) accord neoclassical economic theory to characterizing the efficacy of alternative tenure institutions would need to end, replaced by an interdisciplinary and multicultural perspective capable of capturing the broadly humanistic purposes of land tenure institutions (Ayano 2018).

Study limitations

Our findings are limited by the dearth of relevant articles providing sufficient detail about the contextual factors that may have affected the outcomes of interest of either PDR or LRD interventions in the four case study countries. Most articles about PDR programs identified through the initial database searches were econometric analyses, which typically provided limited information about the context in which the intervention took place. Few of the econometric studies included qualitative data which might have provided explanations for unexpected findings, such as the decline in crop yields on plots held by female heads of households in PFR villages in Benin. In one study (WBGIL 2019), qualitative data were collected as part of the program evaluation, but not incorporated into the findings. The dearth of studies limited not only the depth of the analysis for the four countries, but the number of countries that could be included in the analysis. Most of the studies with extensive discussion of contextual factors were identified through purposive searches, demonstrating the advantage of using a methodology which encourages reviewers to use a range of approaches to identifying relevant material for inclusion in the synthesis. A pressing need exists for program evaluators to incorporate robust ethnographic and other forms of qualitative data collection and analyses into their methodological toolkit with the same level of consideration given to econometric analyses.

References

Ali, D. A., Deininger, K., Mahofa, G., & Nyakulama, R. (2019). Sustaining land registration benefits by addressing the challenges of reversion to informality in Rwanda. *Land Use Policy*, 104317. https://doi.org/10.1016/j.landusepol.2019.104317.

Ayano, M. F. (2018). Rural land registration in Ethiopia: myths and realities. *Law & Society Review*, *52*(4), 1060–1097.

Boone, C. (2007). Property and constitutional order: land tenure reform and the future of the African state. *African Affairs*, *106*(425), 557–586. https://doi.org/10.1093/afraf/adm059.

Butler, M. (2021). Analyzing community forest enterprises in the Maya Biosphere Reserve using a modified capitals framework. *World Development*, *140*, 105284. https://doi.org/10.1016/j.worlddev.2020.105284.

Deininger, K., Savastano, S., & Xia, F. (2017). Smallholders' land access in Sub-Saharan Africa: a new landscape? *Agriculture in Africa – Telling Myths from Facts*, *67*, 78–92. https://doi.org/10.1016/j.foodpol.2016.09.012.

Fitzpatrick, D. (2006). *Evolution and Chaos in Property Rights Systems: The Third World Tragedy of Contested Access* (SSRN Scholarly Paper No. 2007662). https://papers.ssrn.com/abstract=2007662.

Honig, L. (2017). Selecting the state or choosing the chief? The political determinants of smallholder land titling. *World Development*, *100*, 94–107. https://doi.org/10.1016/j.worlddev.2017.07.028.

Knight, R. S. (2010). Statutory recognition of customary land rights in Africa: an investigation into best practices for lawmaking and implementation. *FAO Legislative Study*, *No. 105*. https://www.cabdirect.org/cabdirect/abstract/20113106768.

Lawry, S. (1993). Transactions in cropland held under customary tenure in Lesotho. In T. J. Bassett and D. E. Crummy (Eds.), *Land in African Agrarian Systems* (pp. 57–74). University of Wisconsin Press.

Lawry, S., Samii, C., Hall, R., Leopold, A., Hornby, D., & Mtero, F. (2017). The impact of land property rights interventions on investment and agricultural productivity in developing countries: a systematic review. *Journal of Development Effectiveness*, *9*(1), 61–81. https://doi.org/10.1080/19439342.2016.1160947.

Mkodzongi, G. (2016). 'I am a paramount chief, this land belongs to my ancestors': the reconfiguration of rural authority after Zimbabwe's land reforms. *Review of*

African Political Economy, *43*(sup1), 99–114. https://doi.org/10.1080/03056244.2015.1085376.

Murray, C. (1981). *Families Divided: The Impact of Migrant Labour in Lesotho* (Vol. 29). Cambridge University Press.

Peters, P. E. (2021). Kinship. In A. H. Akram-Lodhi, K. Dietz, B. Engels, & B. McKay (Eds.), *Handbook of Critical Agrarian Studies* (pp. 139–149). Edward Elgar Publishing. https://www.elgaronline.com/display/edcoll/9781788972451/9781788972451.00024.xml.

Scoones, I., Marongwe, N., Mavedzenge, B., Murimbarimba, F., Mahenehene, J., & Sukume, C. (2011). Zimbabwe's land reform: challenging the myths. *The Journal of Peasant Studies*, *38*(5), 967–993. https://doi.org/10.1080/03066150.2011.622042.

Ubink, J. (2011). *Stating the Customary: An Innovative Approach to the Locally Legitimate Recording of Customary Justice in Namibia* (Working Papers Series No. 8; Traditional Justice: Practitioners' Perspectives). International Law Development Organization. https://www.academia.edu/64541194/Stating_the_Customary_An_Innovative_Approach_to_the_Locally_Legitimate_Recording_of_Customary_Justice_in_Namibia.

World Bank Gender and Innovation Lab (WBGIL). (2019). *Impact Evaluation of Access to Land Project in Benin.* Prepared for the Millennium Challenge Corporation. https://thedocs.worldbank.org/en/doc/537351555943343180-0010022019/original/MCCEvaluationReportIEofAccesstoLandProjectinBeninFINAL.pdf.

Annexes

Annex 1—Search terms

Search terms, web of science	
Variable	*Term*
Interventions	Title = (Land OR tenure OR propert* OR Reform OR Security OR Capacity OR Administration OR Conflict OR Rights OR Awareness OR literacy OR document* OR demarcat* OR Use OR planning OR Management OR governance OR collectiv* OR Formal* OR joint* OR titl* OR behavioral OR regulariz* OR justice) AND
Outcomes	Title = (productiv* OR output* OR yield* OR income* OR agricultur* OR invest* OR equit* OR "social inclusion" OR resilienc* OR "adaptive capacity" OR ("Social integration") OR ("Social justice") OR Equality OR Inequality OR Fair* OR ("collective action") OR ("social network") OR Mobility OR Innovative OR Transformative OR Livelihood* OR divers* OR Biodiversity OR Vulnerability OR Shock OR ("Food security") OR ("Vegetation cover") OR ("Forest cover") OR (rent* AND market*) OR (Market AND pric*) OR Diversification OR Efficiency)
3ie keyword search terms	
Keywords	Keywords: Land OR Formal Land Right OR Informal Land Right OR Land Administration OR Land Allocation OR Land Market Restrictions OR Common Property OR Fishery Management OR Common-Property Forestry OR Insecure Property Rights OR Property Regimes OR Property Right OR Property Rights OR Land Tenure OR Land Tenure Contracts OR Land Tenure Security OR Secure-Tenure OR Security Of Tenure Effect OR Tenure Formalization OR Tenure Insecurity OR Tenure Security OR Customary Leaders OR Certification OR Certification Schemes OR Land Title OR Land Titling

Annex 2—Appraisal form

Email address

Record number

Authors

Year of publication

Intervention (as authors label it):

Country/Countries/Sub-continent or continent

Short description of the intervention in your own words

Please, choose the alternative that better fits to describe the methodology in relation to the type of data used: qualitative, quantitative, mixed methods (choose one)

Please answer the following:

- Do the authors discuss gender or analyze data by sex? (yes/no)
- Does the article describe the intervention? (yes/no)
- Does the article study effects on productivity? (yes/no)
- Does the article study effects on investment? (yes/no)
- Does the article study effects on social inclusion? (yes/no)
- Does the article study effects on resilience? (yes/no)
- If the article discusses effects on other development outcomes, please list here. Write none if appropriate
- What are the article's strengths?
- What are the article's weaknesses?
- Does the paper describe the connection(s) between the outcomes and the process (C + M = O)?
- Does this paper need more information about the context?

Final Rating: High, Medium, or Low (please choose one)

High: *means that the paper meets all of the following criteria: (1) the framing of the research and the research questions and outcomes are highly matched to the review questions; (2) the paper studies an intervention or reform; (3) the empirical findings are clearly described; and (4) there is a rich description of the process/mechanism and context that can greatly advance the theoretical output of the review. The paper is a "key informant."*

Moderate: *means that (1) the framing of the research and the research questions and outcomes are highly matched to the review questions, and (2) the paper studies an intervention or reform; however, the paper has one or two of the following drawbacks: (a) lacks relevant empirical data; (b) is relatively thin on the description of context and mechanism; or (c) reports on a different (but related) intervention working toward similar outcomes of interest.*

Low: *It is not placed in the exclusion category because it contains at least one idea or statement about the context, about the mechanisms or about* *conceptualizing outcomes that can be used for refining the theory and building a CMO configuration. It will not be given a full appraisal but may be utilized as a reference.*

***Exclude**: It does not meet criteria 1 and 2 (from above); or it has all of the drawbacks (a, b, and c); or it has serious red flags (e.g. obvious quality issues, incorrect information, poorly written).*

For moderate or low papers, note if study possesses the following:

a. Empirical Data, uses Experimental or Quasi-Exp Methodology
b. Tenure or Non-tenure intervention
c. Enumerates Theory

Comment field (please note the following):

a. The "primary outcome(s) of interest" (if not primarily one of ours)
b. If the study demonstrates "other things matter more" (e.g. besides tenure) for our outcomes
c. If it is rated low (reference only), name possible reference use (e.g. talks about fallowing, etc.)
d. If excluded, why it was excluded

Annex 3—Quality assessment criteria and scoring

Criteria	*Point assignment*	*Scoring*
Conceptual framing • Acknowledges existing research or theory • Clearly situates the study design and analysis in existing body of theory • Constructs or uses an existing conceptual or theoretical framework • Poses specific research questions and may investigate specific hypotheses	Attach 1 point for each yes answer	How well does the study frame the research? 0 = not at all 4 = very well
Transparency: • Clearly explains its design, methods, data collected and analyzed, and other details to allow potential replicability • Identifies funding used for a study and discusses conflicts of interest	Attach 1 point each for: • Clear design/ methods • Clear data collection process • Identifies funding • Identifies potential conflict of interests	How transparent is the study? 0 = not at all 4 = highly transparent

(*Continued*)

Criteria	*Point assignment*	*Scoring*
Appropriateness • Uses appropriate designs and methods in relation to the stated study objectives, hypothesis, and research questions in the study • Reflects on cultural sensitivity and demonstrates that they have taken adequate steps to consider the effect of local cultural dynamics on their research or on a development intervention	Attach 1 point for: • Appropriate design/ methods • Appropriate results interpretation • Reflects cultural sensitivities • Reflects on ethical considerations	How appropriate is the study? 0 = not at all 4 = highly appropriate
Validity • Explains and uses indicators that are well suited for the concept, or variable, under measurement • Explains and uses methods that appropriately minimize the possibility of spurious relations • Explains and uses methods that can be appropriately replicated across multiple contexts • Explicitly considers how far the research findings may have been biased by the activity of doing research itself	• Attach 1 point for each yes answer	How valid is the study? 0 = not at all 4 = high attention to validity
Reliability • Explains how appropriate the instrument of measurement is for the specific variable • Discusses internal reliability in reference to how appropriate measures used are in relation to the cultural context • Discusses how the application of a different analytical technique to the same set of data could produce different results	Attach 1 point for each yes answer	How much attention has the study paid to reliability? 0 = not at all 3 = highly reliable
Cogency • Uses a clear, logical thread that link the theoretical framework to the data, analysis, and conclusions • Is written clearly • Avoids claims not clearly backed up by the data and findings • Identifies limitations	Attach 1 point for each yes answer	How coherent is the study? 0 = not at all 4 = highly coherent

Annex 4—Data extraction form

Intervention: ______________ Country: ____________

Article short citation	
Short description of the intervention	
Outcomes studied (investment, productivity, social inclusion, or resilience)	
Does/do the paper/s describe CMO?	
Do the authors include significant gender analysis?[a]	
Is analysis of vulnerable groups other than women included?[b]	
The intervention	
Quality of the papers (use the score card system 1 = poor quality, 4 = high quality)	
Main problems that the intervention aims to "solve"	*This refers to the problem statements according to the authors of the intervention.*
Definition/measure of tenure security used by studies	*Note: studies hardly ever define tenure security in the same way.*
How did the definition of the problem affect understanding of the problems experienced by women and other vulnerable groups?	*Think about whether the conceptualization of the problem affected the inclusion of women or other groups, or if it included women's needs/interests/other gender considerations.*
Coverage (national, regional, sub-regional)	
Target population (where stated) (e.g., families or women & men, specific groups by age, or ethnicity, status, sex, marital status)	
Main intervention components/areas/activities	
Duration/timeframe of the intervention (dates, e.g., 1990–1995)	

(*Continued*)

Expected theory of change (if graphic is presented, paste at end of form)	
Mechanisms	*Hypothesis of the intervention. Explain the link between the problem, inputs and how the intervention is expected to affect outcomes.*
How did the intervention's theory of change and mechanisms deal with gender?	*Please think about this beyond sex disaggregation*
Intermediate outcomes leading to…	Productivity/investment
	Social inclusion
	Resilience
Did the intervention designers think about considerations to include women and other vulnerable groups or to address gender social norms?	
Final productivity & investment outcomes	
Final inclusion outcomes	
Final Resilience outcomes	
How was gender included in the expectations about the effects of the interventions on all the final outcomes?	
How do the interventions align with an understanding of the context?	
How is the problem defined in the context?	*The problem as perceived by you after reading contextual information that reflects on the problems as identified by the beneficiaries or the authors of the papers about the context.*
Household Production Orientation: to what degree are HHs oriented toward dedicating their land and labor principally to agricultural production, for market and/or home consumption? How significant are non-agricultural uses of HH land and labor as a proportion of the total allocation of assets? Did the assumptions underlying the intervention incorporate an accurate reflection of the % of agriculture as a source of HH income?	

How did people access land at the time of the intervention, i.e., what was the general type of tenure system? (1) Customary (access to land as a social or inherited right) (2) purchase, lease, or inheritance of title; or (3) other (including state land allocation) *Did this differ by sub-population (e.g., women, youth)?*	
Did the intervention change the nature of the system (e.g., from customary to private or freehold land, or from state to statutorily recognized tenure) or rather did it seek to strengthen tenure security within an existing system (e.g., through certification of existing customary or private rights?)	
Describe the bundle of rights before and after the intervention.	*Could you sell it? Bequeath it? Rent it? Decide what to grow? What to do with the profits?*
What were the land rights of women and vulnerable groups before and after the intervention?	
Provide a summary discussion of the success of the intervention in clarifying or strengthening tenure security, and on the degree to which the intervention extended or reduced inclusion of women/vulnerable groups as bona fide rights holders (on par or more equal with men/more privileged)	*Did the hypothesis hold? Why did that happen? For example if the hypothesis assumed that the households had an agricultural orientation, do the studies about the context support that?*
Unintended negative or positive consequences of the intervention	
Contextual factors relevant to understanding failure or success of the intervention, taking into account how different groups experienced outcomes differently	
CMO	*A summary comparison of the program theory of change and what happened, pointing out the gaps*
Comments	

[a] Meaning purposeful consideration with some reasoning, explanation or theory behind the gender analysis.
[b] Meaning purposeful consideration with some reasoning, explanation or theory behind the analysis of vulnerable groups.

Annex 5a—Parcel demarcation and rights registration—tenure security outcomes

Benin		*Ethiopia*		*Rwanda*	
(+)*	PFR villagers more likely to have clearly demarcated parcel boundaries (Goldstein et al. 2018; WBGIL 2019)	(+)	FLLC reduced fears of land loss in Amhara (Deininger et al. 2011; Melesse and Bulte 2015; Yami and Snyder 2015)	(+)	Tenure security increased overall (Abbott et al. 2018), but more for women (Ali et al. 2014)
(/)	No impact on incidence of land conflicts (Goldstein et al. 2018; WBGIL 2019; Yemadje et al. 2014)	(+)	FLLC led to reduction in land conflicts (Kumar and Quisumbing 2015; Melesse and Bulte 2015; Yami and Synder 2015) or perceptions that are less likely to have conflict (Ahmed 2017)	(+) (−)	Boundary disputes decreased but disputes over land inheritance increased (Muyombano et al. 2018)
(+)*	PFR villages had 2% fewer intra-village land conflicts than control villages [*but incidence of conflict in control villages very low*] (Wren-Lewis et al. 2020)	(+)	FLLC led to increase in renting out (Holden et al. 2011; Yami and Snyder 2015)	(−)	Children whose parents do not have legally registered marriages will be ineligible to inherit land (Muyombano et al. 2018)
(+)	PFR villages had 13.5% fewer inter-village land conflicts (Wren-Lewis et al. 2020)	(−)	FLCC holders still have fears of state expropriation (Yami and Snyder 2015)	(/)	No impact on perceptions of risk of state expropriation (Ali et al. 2014)
(+)*	Average percentage of respondents who believed local land governance institutions could resolve land conflicts higher in PFR villages (Wren-Lewis et al. 2020) [*but belief in control villages that local institutions could resolve conflicts relatively high*]	(+)	Increase in perception that will be compensated if state expropriates land (Kumar and Quisumbing 2015)	(Unclear)	Landholders perceived that LTR made it more likely land disputes would be resolved and that the number of disputes had declined, but no evidence as to whether this reflects actual practice (Abbott and Mugisha 2015)
(−)	PFR exacerbated tensions and led to conflicts between tenants and landowners in the Adja Plateau (Yemadje et al. 2014)	(+)	Most (81%) of landlords reported greater bargaining power with tenants, fewer land disputes with tenants (Holden et al. 2011)	(+)	Landholders perceived that LTR made it more likely that land disputes would be resolved (Abbott et al. 2018)

(−)	PFR catalyzed conflicts between founding lineage members and long-term in-migrants in Collines Department (Lavigne Delville and Moalic 2019)	(+)	Heads of households that have land certificates more likely to perceive that they have tenure security (Ahmed 2017; Melesse and Bulte 2015)	(+)	Uncertainty about inheritances reduced (Ali et al. 2014)
(−)	PFR catalyzed conflicts between founding villages and satellite villages/hamlets (Lavigne Delville and Moalic 2019)	(+)	Heads of households that have land certificates more likely to have at least one household member migrate, which authors interpret as sign of tenure security (de Brauw and Mueller 2012) [but caveat is that relationship is weak]	(−)	Landowners, especially near urban areas, still displeased with state expropriation practices (Abbott and Mugisha 2015)
(+)	Parcels in PFR villages more likely to have written documentation of land rights (WBGIL 2019) or for tenancy agreements (Yemadje et al. 2014)	(/)	No difference in concerns about land disputes between those with FLLC and those without (Deininger et al. 2011)	(/)	No evidence of increased feelings of tenure security (Abbott and Mugisha 2015)
(/)	No impact on whether landowners had written documentation of their rights (Yemadje et al. 2014)	(/)	No difference in time to resolve land disputes with SLLC versus FLLC (Cloudburst Group 2016)		
		(+)	SLLC recipients more likely than FLLC recipients to state that they had heritable right to bequeath land (Cloudburst 2016)		
		(−)	SLLC participants still have fears of state expropriation (Cloudburst 2016)		
		(/)	No difference in land conflicts with SLLC vs FLLC (Cloudburst Group 2016)		

(+) = positive outcome; (−) = negative outcome; (/) = neutral outcome; * indicates caveat for outcome category.

Annex 5b—Parcel demarcation and rights registration—investment outcomes

Benin		*Ethiopia*		*Rwanda*	
(+)	Increase in tree-planting after one year and four years (WBGIL 2019)	(+)	Increased investment in soil and water conservation structures (Ahmed 2017; Deininger et al. 2011; Melesse and Bulte 2015)	(+)	Increased investments in soil conservation structures (Ali et al. 2014; Bayisenge 2018)
(+)	Minor increase in planting of perennial crops over short (one year) and long term (four years) (WBGIL 2019)	(/)	No impact on investment from second level certification (Cloudburst 2016)	(/)	No impact on amount of land with soil erosion protection structures (Abbott and Mugisha 2015)
(+)	Increased likelihood that female heads of household leave demarcated fields in fallow (WBGIL). Evidence as to why they were more likely to leave demarcated lands in fallow was not provided	(+)	Increased tree-planting (Melesse and Bulte 2015)	(/)	Observed increase in use of soil fertility enhancement inputs attributed to subsidized inputs made available through land consolidation program to households that have land certificates
(+)	Reduced tree cover loss and lower incidence of fires (Wren-Lewis et al. 2020)	(+)	Increase in use of organic fertilizer (Melesse and Bulte 2015)	(/)	Respondents did not perceive that LTR had impacted investments in land improvements (Abbott and Mugisha 2015)
(+)	Increased use of organic fertilizers by landowners and tenant farmers (Yemadje et al. 2014)				
(+)	Increased use of mineral fertilizers by landowners (Yemadje et al. 2014)				
(−)	Decline in oil palm intercropping (a traditional method for improving soil fertility) (Yemadje et al. 2014)				

(+) = positive outcome; (−) = negative outcome; (/) = neutral outcome; * indicates caveat for outcome category.

Annex 5c—Parcel demarcation and rights registration—agricultural productivity, land market, and credit outcomes

Benin		*Ethiopia*		*Rwanda*	
Agricultural productivity					
(/)	No impact on agricultural productivity (Goldstein 2015, 2018; WBGIL 2019)	(+)	Productivity gains worth $US 75.40 (higher on owned plots than rented plots) (Melesse and Bulte 2015)	(/)	No impact on rate of increase in agricultural productivity (Abbott and Mugisha 2015)
(/)	No impact on off-farm employment (WBGIL 2019)	(+)	Shift toward greater crop diversification and to crops requiring more labor (Yami and Snyder 2015)		
(−)	Reduction in yields for female heads of household by 20% (WBGIL 2019)	(+)	Not directly measured but argues that increase in rentals likely translates into productivity gains (Deininger et al. 2011; Holden et al. 2011)		
Land markets					
(−)	Reduction in land rentals/sharecropping in PFR villages (Goldstein et al. 2015)	(/)	No impact of Second Level Land Certification on land rentals (Cloudburst Group 2016)	(/)	No impact on land sales (Abbott and Mugisha 2015; Muyombano et al. 2018)
(+)	More plots rented out in PFR villages (Yemadje et al. 2012, 2014) (but lacks evidence of causality because study baseline data was not collected)	(+)	Increase in land rented out (especially by female heads of household) (Deininger et al. 2011; Holden et al. 2011; Yami and Snyder 2015)	(−)	Decrease in land sales (Ali et al. 2014)
				(+)	Increase in rentals (Muyombano et al. 2018)
				(+)	Land values increased (Muyombano et al. 2018)

(*Continued*)

Credit effects

None of the Benin studies discussed credit effects	(/)	No impact on use of land certificates for acquiring credit (Yami and Snyder 2015)	(/)	12% of LTR recipients used their titles to get loans; for those who farmed their own land, one-third used loans for agricultural activities; for those who did not farm their land, 10% used the loans for agricultural activities (Abbott and Mugisha 2015)
	(Unclear)	Some increase in use of credit by those with second level land certificates but unclear what source of credit was, and credit typically not used for investing in agriculture (Cloudburst Group 2016)	(/)	No credit effect (Ali et al. 2014)
			(+)	Younger and wealthier farmers more likely to use certificate as collateral; however, loans often were not used for agricultural investments (Muyombano et al. 2018)

(+) = positive outcome; (−) = negative outcome; (/) = neutral outcome.

Annex 5d—Parcel demarcation and rights registration—social inclusion outcomes

Benin		Ethiopia		Rwanda	
Gender inclusion					
(−)	Tenure security weakened for many women (Giovarelli et al. 2015)	(+)	Female heads of households more likely to report perceiving greater tenure security from certificates than male heads of households (Ahmed 2017)	(+)	60% of land jointly registered, including nearly half of women in non-registered marriages (Abbott 2018)
(+)	In treatment villages, female heads of household more likely to leave demarcated land in fallow (Goldstein et al. 2018)	(/)	No difference between men and women in perceptions of tenure security (Deininger et al. 2011)	(+)	Women more able to claim their land rights (Abbott and Mugisha 2015)
(−)	Female heads of household had roughly 20% lower yields than male heads of households (WBGIL 2019)	(+)	Landlords (many of whom were female heads of household) reported land certificates provided greater tenure security (Deininger et al. 2011; Holden et al. 2011)	(+)	Women have better understanding of inheritance and ownership rights (Abbott and Mugisha 2015)
		(+)	Female heads of households with land certificates more likely to rent out land (Deininger et al. 2011)	(+)	Female headed households and women in legally registered marriages more likely to have documented land ownership (Ali et al. 2014)
Migrant inclusion (data available only for Benin)		(−)	Women less likely to have registered land (Ahmed 2017; Kumar and Quisumbing 2015)	(+)	Female headed households and women in legally registered marriages more likely to invest in conservation structures (Ali et al. 2014)
(−)	Conflicts catalyzed between in-migrants and founding lineage members (Lavigne Delville and Moalic 2019)	(+) (−)	Requirement for joint titling requirement provided married women with greater tenure security but effect is mediated by cultural norms (Lavers 2017)	(+) (−)	Tenure security increased for women in formally registered marriages but reduced for those not in registered marriages (Bayisenge 2018; Muyombano et al. 2018)
(−)	Conflicts catalyzed between hamlets and founding villages (Lavigne Delville and Moalic 2019)	(−)	Female heads of household less aware of the land certificate program than male heads of household (Ahmed 2017; Kumar and Quisumbung 2015)	(+)	Women had greater role in decision-making post-LTR (Bayisenge 2018)

(*Continued*)

Benin		*Ethiopia*		*Rwanda*	
(−)	Some tenants reported losing access to land (Yemadje et al. 2012, 2014)	(+)	Land certificate improves women's status (Kumar and Quisumbing 2015)	(−)	Tenure security reduced for co-wives in polygamous marriages (Muyombano et al. 2018)
		(+)	Joint land titling positively associated with greater access to land for women, rentals by women, and decision-making for crops (Yami and Synder 2015)	(+)	Women with plots titled in their name only had increased tenure security (Santos et al. 2014)
		(+)	Increased likelihood that wife makes decisions about which crops to grow on her land if have SLLC compared with FLLC (Cloudburst Group 2016)	(+)	Women in polygamous marriages more likely to have plots titled in their name and more likely to perceive gains in tenure security for such plots, including having their children registered as heirs (Santos et al. 2014)
		(+)	Increased likelihood that wife has land in her own name if have SLLC compared with FLLC (Cloudburst Group 2016)	(+) (−)	Women in registered marriages more likely to benefit from joint titling but customary norms weaken their tenure security (Santos et al. 2014)
				(−)	Poor women less likely to be included on joint titles (Santos et al. 2014)
				(−)	Daughters in poorer households much less likely to inherit a plot than daughters in better off households (Santos et al. 2014)

Annex 6a—Contextual factors affecting outcomes in Benin

Study	*Contextual factors identified as impacting PDR outcomes of interest in Benin*
Giovarelli et al. (2015)	**Affecting tenure security and social inclusion (gender) negatively**
	Cultural norms in which women have weaker rights to land than men
	Weak negotiating position of women
	Lack of understanding on part of policymakers and program designers of complexities and normative goals of customary tenure systems
	Project implementation factors affecting social inclusion negatively
	Program timeframe too short for the task of recording secondary rights
	Program teams inadequately trained/supervised in how to record secondary rights
Goldstein et al. (2015/2018); WBGIL (2019)	**Potentially affecting tenure security negatively (for those who feel certificates provide greater security)**
	Weak level of commitment of state officials and village land commissions to land certificate delivery (WBGIL 2019)
	High cost of certificates (both fees and transaction costs) identified as a barrier to individuals obtaining certificates (WBGIL 2019)
	Affecting social inclusion negatively
	Members of land committees and village advisers more likely to have a land certificate than others (WBGIL 2019)
	Affecting social inclusion (gender) outcomes positively
	Cultural norms in which women have weaker rights to land than men led to an increase by female headed households in letting land that had been demarcated lie fallow (Goldstein et al. 2018; WBGIL 2019)
Lavigne Delville and Moalic (2019)	**Affecting tenure security outcomes/social inclusion (migrants) negatively**
	Frontier area with large numbers of long-term in-migrants with access to land via "tutorat" system was a context ripe for conflict over land claims
	Founding villages allocate land to satellite villages which led to conflicts over which village controlled the territory being mapped by PFR
	In-migrants can gradually acquire strong rights under customary system norms; this led to conflicts when founding lineages sought to record rights in their name
	Policymakers/program managers had inadequate knowledge of how customary rights systems work
	Affecting tenure security outcomes/social inclusion (migrants) positively
	Stronger political position of administrative villages enabled them to successfully assert claims under PFR
	Project implementation factors affecting social inclusion (women/migrants) negatively
	Project timeframe too short for the task of recording secondary rights
	Project teams inadequately trained in how to record secondary rights (use of persons with land survey background rather than social scientists)

(*Continued*)

Study	*Contextual factors identified as impacting PDR outcomes of interest in Benin*
Wren-Lewis et al. (2020) Yemadje et al. (2012, 2014)	**Affecting tenure security and deforestation outcomes positively:** Village-level land governance system largely functional in both PFR and non-PFR villages **Affecting tenure security outcomes negatively** High proportion of tenancy arrangements; most of which were previously undocumented State incapacity to deal with land conflicts **Affecting social inclusion outcomes (tenants) negatively** Power balance biased in favor of landlords **Affecting tenure security outcomes/interest in participating in program positively** Relatively high population density High demand for land suitable for oil palm production (commodity crop) Customary authority system still fairly strong but many perceive state-approved documents as having more weight

Annex 6b—Contextual factors affecting outcomes in Ethiopia

Study	*Contextual factors identified as impacting PDR outcomes of interest in Ethiopia*
Ahmed (2017)	**Affecting tenure security outcomes/interest in participating in program positively** History of state redistribution of land and some likelihood in most regions that it might occur again in future increases interest in certificates **Affecting social inclusion negatively** For men, landholders further from roads less likely to have a land certificate For men, poorer landholders less likely to have land certificates Women with lower social status and weaker family support are less able to protect their access to land Women culturally not viewed as "key farming agents" despite constitution half the agricultural labor force In Amhara, most landholders are men (so women are less likely to get land certificates) In practice, few women are on land committees despite requirement that they be included **Affecting social inclusion (tenure security for women) outcomes positively:** 1995 Constitution specifies equal rights for men and women for land access, management, transfer, inheritance Land policy requires joint titling for married couples; wife has to countersign transfers **Affecting women's participation in land markets positively** Women (especially female heads of households) often sharecrop out their land because they lack access to key farming resources (oxen, plows); this puts them at risk of losing their land (because others are farming it); having a land certificate reduces this risk
de Brauw and Mueller (2012)	**Affecting tenure security outcomes positively** Land belongs to the state and landholders have use rights only; this coupled with history of expropriation by the state led to fears that land might be taken for redistribution; having a land certificate has reduced those fears Households with a member who has out-migrated are more likely to feel their tenure is secure **Affecting land market outcomes** Rentals are permitted but limitations exist on the amounts that can be rented out and duration of contracts Landholders cannot sell, mortgage, or exchange land
Cloudburst Group (2016)	**Affecting tenure security outcomes** At baseline, first-level certification had already been done and most household heads perceived risk of land expropriation was low **Affecting overall outcomes positively** Program was implemented in areas with high agricultural potential, good access to markets, presence of agricultural investors, and high level of land transactions (rentals/sharecropping) Positive impacts more likely in places close to urban areas

(*Continued*)

Study	*Contextual factors identified as impacting PDR outcomes of interest in Ethiopia*
Deininger et al. (2011)	**Contributing to overall success of program**
	Initial certification done with minimal donor investment; as a result, approach adopted was low-cost and easily accessible to most landholders
	Affecting interest in participating in program positively
	Demand for certificates higher near urban areas
	Perceived tenure insecurity very high prior to intervention due to recent land redistribution
	Extensive awareness-raising efforts reached women and poor as much as men and wealthier individuals
	Wealthier and older heads of households more likely to fear land loss (due to government redistribution) than others (attributes this to land governance system that seeks equitable land distribution among residents)
	Affecting social inclusion (tenure security for women) positively
	Promotion of joint titling in later phases
	Affecting land markets/social inclusion
	Cultural norms against women using oxen makes it difficult for them to farm their land but there are risks to renting out land
	Increase in rental markets as possession of certificate is perceived to reduce risks associated with renting; result is a transfer of land (via rental agreements) from households lacking oxen (usually female headed households) to households with oxen
Holden et al. (2011)	**Affecting overall success positively**
	Local participation in implementation of early pilots ensured a sense of "ownership" by local government
	Land certification program emphasized working through local governance institutions that were well-established
	Low cost and wide coverage avoided elite capture
	Reform did not threaten elites; had transparent process
	Affecting tenure security outcomes/interest in program positively
	Land belongs to state; fears of expropriation encouraged interest in obtaining land certificates
	Affecting social inclusion (tenure security for women) positively
	Female headed households received certificates in their own names (enhancing their security)
	Affecting social inclusion (tenure security for women) negatively
	Cultural norms in which women move to husband's when marry places them in a weak bargaining position (thereby making it challenging to request joint title and therefore not enhancing their tenure security)
	Affecting land markets
	Only men can use oxen; this restricts female-headed households' ability to farm; at the same time, renting out the land puts them at risk of having the local government redistribute the land or in-laws may expropriate it; having a land certificate reduces these risks
	Shortage of rental land available makes renting out potentially advantageous

Kumar and Quisumbing (2015)	**Affecting tenure security outcomes/interest in participating in program positively** Every Ethiopian has a right to land and administrative reallocations are always possible (creating incentive for acquiring land certificate) **Affecting interest in participating in program negatively** In areas where plots clearly demarcated prior to intervention, certificate less likely to be viewed as important for protection from encroachment **Affecting social inclusion (tenure security for women) outcomes positively** Revised Family Code in 2000 provided equal division of assets on divorce Regions where program was active other than Tigray initially all required joint certification In Tigray, despite not having joint certification requirement initially, overall women had stronger rights in general **Affecting social inclusion (tenure security for women) outcomes** The status of women varies across the country by religion and ethnic affiliation, but generally social norms favor men in property rights and in bargaining power Female heads of household tend to have fewer assets, less social capital, less education, less access to financial capital, all of which tend to put them in a weak bargaining position vis-à-vis others
Lavers (2017)	**Affecting tenure security outcomes/interest in participating in program positively** Land reform done in context of previous land redistribution which had created insecurity **Affecting social inclusion (tenure security for women) outcomes positively** State legal framework supportive of equal rights in marriage, inheritance, and land Federal Land Proclamation require joint titles for husband and wife **Affecting social inclusion (tenure security for women) outcomes** Nature of interaction between state institutions and customary norms as well as power relations within households are important factors in whether women can actualize their land rights Tigray had social norms of gender equality and under the TPLF, women's land rights were emphasized and women effectively had access to at least some land in their own right In Oromia, social norms are less favorable for women's land rights than in Tigray. Even though women may have their names on joint titles, their ability to prevail in land disputes is limited since land disputes are resolved at the local level where local social norms prevail households

(*Continued*)

Study	*Contextual factors identified as impacting PDR outcomes of interest in Ethiopia*
Melesse and Bulte (2015)	**Affecting tenure security outcomes/overall interest in participating in program positively**
	History of redistribution means that state landownership is a source of insecurity
	Customary land tenure systems less important than elsewhere in Africa
	High population density area with good conditions for agriculture and marketing potential
	Various sources of uncertainty include urban expansion, land grants to investors, opaque justice system
	Affecting social inclusion outcomes (tenure security for women) positively
	Program support/requirements for joint titling for spouses
Yami and Snyder (2015)	**Affecting tenure security outcomes/interest in participating in program positively**
	History of forced land redistribution created fear of expropriation especially in Amhara and SNNP Region; land redistribution seen less as a threat in Tigray sites; fears have diminished with certification
	Affecting social inclusion (tenure security for poorer villagers and women) outcomes negatively
	Information on landholdings often collected from household heads, married women sometimes excluded
	Women not represented on the land committees despite guidelines that they should be due to issue of cultural norms where women are not considered to be capable of resolving conflicts related to land
	Social norms in which women have weak land rights and limited voice in decisions at household and village level
	Elites controlled land registration and some poorer villagers were left out when high value irrigated land was redistributed
	Affecting sustainability of program negatively
	Limited capacity of land administration committees
	Lack of commitment from politicians

Annex 6c—Contextual factors affecting outcomes in Rwanda

Study	*Contextual factors identified as impacting PDR outcomes of interest in Rwanda*
Abbott et al. (2018)	**Affecting tenure security outcomes positively** Land belongs to the state (users have 99-year leases) which has engendered fears of state expropriation, especially near urban areas; certificates are viewed as providing some protection **Affecting social inclusion outcomes positively** Rwanda has enacted policies/laws to eliminate gender inequalities including through inheritance and land laws, and emphasizing rights for married women **Affecting social inclusion negatively** Women's lack of ability to enforce their rights when go to court over land claims Cultural norms that give women have lesser status than men especially regarding land decisions **Project implementation factors affecting social inclusion negatively** Cohabiting partners can register jointly under LTR (and some do) but project teams assumed that only women in registered marriages could get joint titles with their husband **Project implementation factors affecting social inclusion positively** Rights for married women required as part of registration
Abbott and Mugisha (2015)	**Affecting social inclusion outcomes positively** Broad awareness of 1999 inheritance law has been a factor enabling women to exercise rights to land **Attenuating positive impacts on social inclusion outcomes** Limited knowledge of details of women's right to land, especially widows' rights to land Limited training on land rights for men and women Limited training of village leaders in state law and women land rights that results in their mediation of land disputes according to customary norms that favor men in conflict resolution **Affecting investments in land (positive for those in land consolidation program; neutral for those not in program)** Subsidized inputs are available only to landholders who participate in Rwanda's land consolidation program **Affecting credit outcomes negatively** Most smallholders in Rwanda refrain from using land as collateral to apply for loans Low availability of financial products that are suitable for small entrepreneurs Most smallholders have limited financial literacy and business skills High interest rates and short repayment periods for loans make credit unattractive

(Continued)

Study	*Contextual factors identified as impacting PDR outcomes of interest in Rwanda*
Ali et al. (2014)	**Affecting social inclusion (tenure security for women) positively** State marriage and land laws supportive of gender equality **Affecting social inclusion (tenure security for women) negatively** Gaps in the law that don't protect women in unregistered marriages **Lack of impact on land markets** Lack of land market impacts linked to fees for registering transfers and to a prohibition on subdividing parcels smaller than 1 ha. [This law has recently changed, see: https://www.newtimes.co.rw/news/five-major-changes-draft-land-law#:~:text=The%20new%20draft%20land%20law%20enforces%20respect%20of, effects%20irrespective%20of%20any%20kind%20of%20prescription.%20editor@newtimesrwanda.com] **Affecting credit outcomes negatively** Authors attribute lack of credit effect to still incomplete formal land transfer system at time of study
Ali et al. (2019)	**Affecting tenure security negatively** Five years after first-time registration 87% of land transactions not registered **Affecting social inclusion negatively** Women and poor may face higher levels of conflict and reduced access to digital services Transactions for housing sales more likely to be registered than for agricultural land, suggesting that households perceive housing a more valuable asset than agricultural land
Bayisenge (2018)	**Affecting social inclusion (tenure security for women) positively** Legal reforms that sought to provide women with equal rights to property by changing land and inheritance laws **Affecting social inclusion negatively** Customary system in which women access property through male relatives or husband that is resistant to change Weak legal protections for women, whether in monogamous or polygamous unions Many low-income women in rural areas have limited knowledge of the land laws Women with less education and who are older are less willing to let their female offspring inherit their land Poorer women are more likely to be in an unregistered marriage Fees for obtaining land certificates are the primary reason that roughly one-quarter of the women in the study had not claimed their land certificates

Muyombano et al. (2018)	**Affecting tenure security outcomes positively:** Recent law reform gave all children (including girls) in state-recognized marriages equal inheritance rights to land **Attenuating positive tenure security outcome:** The law reform giving equal inheritance rights for male and female offspring did not include children of co-wives in polygamous marriages Most women had access to land through husbands or a male relative; customary norms still commonly applied **Affecting credit outcomes negatively:** Large number of plots are less than 1 ha; limits interest in using land for collateral as not certain will produce enough to pay off loan **Affecting credit outcomes positively:** Wealthier farmers in farmer associations were more likely than others to be able to use land use rights as collateral **Inhibiting land markets:** Cultural norms against land sales still strong Land transfer process costly and time-consuming
Santos et al. (2014)	**Affecting social inclusion (tenure security for women) outcomes positively:** Marriage and land law support gender equality in land rights **Attenuating positive social inclusion outcomes:** Women are more likely than men to gain rights to land through marriage or inheritance Marriage law recognizes only monogamous civil marriages, leaving women in polygamous households and those who are cohabiting with no formal rights to land (unless they register land in their own right) **Variable impact on social inclusion outcomes:** Intrafamily dynamics are primary factors affecting women's tenure security **Project implementation factors affecting social inclusion negatively:** Mistranslation of guide may have affected local officials' and communities' understanding of land rights of women who were not legally married and their children

Annex 7—Zimbabwe's fast track land redistribution program: outcomes and contextual factors

Study	*Outcomes*				*Contextual factors*
	Tenure security	*Conservation investments*	*Productivity*	*Social inclusion—gender*	
Zikhali and Chilonda (2012)	FTLRP beneficiaries felt 99-year lease documents provided secure tenure	Not addressed	Higher productivity by FTLRP beneficiaries compared with communal land farmers, but difference due to subsidies for FTLRP; Productivity of FTLRP half that of white commercial farmers	Not addressed	History of colonial land disposition and post-independence demands for land redistribution, including from influential war veterans Land reform beneficiaries lived in tribal reserves, where customary tenure was secure FTLRP made five million hectares of commercial farming land available for resettlement, included limited basic infrastructure and farmer support services, such as fertilizer Fertilizer subsidies not available for communal farms Labor use more intensive on communal farms; with somewhat higher percentage of female headed households than FTLRP farms
Zikhali (2010)	Tenure insecurity higher on FTLRP land than in communal areas (based on soil conservation investments as indicator of security) FTLRP allotment holders perceive tenure as unstable in medium to long term	Greater investment in communal areas for contour earthen ridges than in FTLRP areas	Not addressed	Not addressed	Land conservation already practiced in communal areas Tenure security on communal farms generally secure FTLRP carried out at an accelerated pace, overriding legal procedures, and catalyzed by land invasions led by war veterans and traditional authorities Policies and aims of FTLRP changed repeatedly in course of implementation Communal households more likely to receive remittances and to be involved in off-farm activities compared to FTLRP households The communal group exhibits stronger social ties (social capital) than the FTLRP group FTLRP beneficiaries had more land but social network diminished after re-located

Marongwe (2011)	Not addressed	Not addressed	Much of FTLRP land not farmed: productivity declined by 58% between 2000 and 2009	Not addressed	The A2 FRLRP distributed white-owned commercial farms to aspiring black commercial farmers Goromonzi district, where the study took place, borders Harare, the capital city Land acquisition process lacked finality—unexplained withdrawal of offer letters and some farms de-listed or dropped from the land acquisition list years after being occupied Criteria for selecting beneficiaries for A2 farms emphasized the potential to use the land productively but were ignored in practice Land allocation institutions were captured by members of the ruling party and by representatives of the state security apparatus Most beneficiaries were drawn from the governing or the local elite Land allocation was deeply influenced by social and political relations of patronage Many beneficiaries lacked capital to invest in commercial agriculture or lacked farming skills
Mutopo (2011)	Access through household negotiations; author says, "this brings some insecurity," but women invested in vegetable gardening nonetheless	Not addressed		Women accessed FTLRP land to farm vegetables which they sold in markets in South Africa	Traditionally, married women gain access to land through negotiations within the framework of the family. This also applied to land secured on new FTLRP farms Families on the Merrivale Farm came from the surrounding communal areas The FTLRP recognized the rights to women, typically unmarried female household heads, to land allocations (but only 18% of those receiving offer letters were women)

(Continued)

Study	*Outcomes*				*Contextual factors*
	Tenure security	*Conservation investments*	*Productivity*	*Social inclusion—gender*	
				Few women able to get their own land on FTLRP area	Traditional authorities and some state officials initially resisted women's rights under FTLRP but resistance eventually overcome Advocacy by women's support groups was important to overcoming resistance The FTLRP presented a life opportunity for most women that had never happened in the history of land relations in Zimbabwe Post-independence politics emphasized social inclusion and racial justice
Scoones (2020)	Tenure insecurity persists across Zimbabwe due to failure 20 years after start of FTLRP to achieve political agreement over how best to provide tenure security in various rural contexts	Not addressed		Not addressed	Lack of investment in administrative and bureaucratic systems necessary to deliver security and clarity Failure to award compensation for commercial farming land confiscated has restricted commercial and donor investment in agriculture. (Agreement on compensation measures may be near) Debate on land tenure policy is ideologically divisive. Needed is an approach appropriate to the circumstances arising in different contexts, for instance, "leases for larger A2 farms, registered permits for new A1 farms, and selective registration for some parts of communal areas, as required"

Index

Note: **Bold** page numbers refer to tables; *italic* page numbers refer to figures.

For Product Safety Concerns and Information please contact our EU representative GPSR@taylorandfrancis.com Taylor & Francis Verlag GmbH, Kaufingerstraße 24, 80331 München, Germany